W9-AFL-548

WITHDRAWN
L. R. COLLEGE LIBRARY

581.3
M95m

143506

| DATE DUE | | | |
|----------|----------|----------|----------|
|          |          |          |          |
|          |          |          |          |
|          |          |          |          |
|          |          |          |          |
|          |          |          |          |
|          |          |          |          |
|          |          |          |          |
|          |          |          |          |
|          |          |          |          |
|          |          |          |          |
|          |          |          |          |
|          |          |          |          |
|          |          |          |          |

# Molecular
# Plant
# Development

# Molecular
# Plant
# Development

**Terence M. Murphy**
*University of California, Davis*

**William F. Thompson**
*North Carolina State University, Raleigh*

CARL A. RUDISILL LIBRARY
LENOIR RHYNE COLLEGE

PRENTICE HALL, Englewood Cliffs, New Jersey 07632

*Library of Congress Cataloging-in-Publication Data*

MURPHY, TERENCE M. (date)
  Molecular plant development.

  Bibliography.
  Includes index.
  1. Plants—Development.   2. Plant molecular genetics.
  I. Thompson, William F. (William Francis) (date)
  II. Title.
  QK731.M87 1988        581.3        87–17403
  ISBN 0-13-599465-9

581.3
M95m
143506
Mar.1988

Editorial/production supervision and
  interior design: Marianne Peters
Cover design: Diane Saxe
Manufacturing buyer: Paula Benevento

© 1988 by Prentice Hall
A Division of Simon & Schuster
Englewood Cliffs, New Jersey 07632

All rights reserved. No part of this book may be
reproduced, in any form or by any means,
without permission in writing from the publisher.

Printed in the United States of America
10  9  8  7  6  5  4  3  2  1

ISBN 0-13-599465-9    01

Prentice-Hall International (UK) Limited, *London*
Prentice-Hall of Australia Pty. Limited, *Sydney*
Prentice-Hall Canada Inc., *Toronto*
Prentice-Hall Hispanoamericana, S. A., *Mexico*
Prentice-Hall of India Private Limited, *New Delhi*
Prentice-Hall of Japan, Inc., *Tokyo*
Simon & Schuster Asia Pte. Ltd., *Singapore*
Editora Prentice-Hall do Brasil, Ltda., *Rio de Janeiro*

# Contents

Preface   ix

## 1.    Developmental Stages in the Plant Life Cycle   1

A.   The Life Cycle of Higher Plants   1

   *Table 1-1: Some Developmental Events in Plant
   Life Cycles.   10*

B.   Classical Concepts of Development   10

C.   Regulators of Growth and Development   15

D.   Molecular Explanations for Development   23

   *Table 1-2: Gene Products Involved in Development   25*

E.   Choosing Experimental Systems   27

## 2.    Gene Expression in Eukaryotes   30

A.   The Prokaryote Model   31

B.   The Eukaryotic System   39

C.   Strategies for Studying Control of Gene Expression   52

# 3.    Detection of Induced Gene Products    54

A.   Heat Shock    54

   Technical Insert: PAGE (Polyacrylamide Gel
   Electrophoresis)    58

B.   The Phenolic Pathway    60

   Technical Insert: In Vitro Protein Synthesis    64

C.   Anaerobiosis and Alcohol Dehydrogenase    65

   Technical Insert: cDNA (Complementary DNA,
   Copy DNA)    72

   Technical Insert: Cloning    73

   Technical Insert: Blot Hybridization    75

D.   Auxin-Induced Gene Products    77

E.   Seed Protein    81

   Technical Insert: Determining the Sequence of DNA    88

F.   Root Nodules and Nitrogen Fixation    91

G.   Chloroplast Proteins    97

H.   Summary    103

# 4.    Organization of the Nuclear Genome and Its Genes    105

A.   Genome Organization    105

   Technical Insert: DNA Reassociation in Solution    120

   Technical Insert: Genomic Cloning    123

B.   Gene Structure    126

# 5.   Genomes of Chloroplasts and Mitochondria   146

A.  Chloroplast DNA   146

   *Table 5-1: Chloroplast Genes That Have Been Mapped on Chloroplast DNA   151*

   *Technical Insert: Restriction Mapping   159*

B.  Mitochondrial Genome   163

   *Table 5-2: Sizes of Some Higher Plant Mitochondrial Genomes   164*

   *Table 5-3: Genes Identified in Plant, Yeast, or Animal Mitochondria   171*


# 6.   Manipulating the Genome   174

A.  Transposons   175

B.  Cauliflower Mosaic Virus   181

C.  *Agrobacterium tumefaciens*   184

D.  Examples of Transformation   189

   *Technical Insert: Electroporation   197*


# 7.   Summary   200

A.  How Genes Are Controlled in Development   201

B.  How Genes Control Development   202


# Bibliography   204

# Index   215

# Preface

The objective of this book is to present to the advanced under-graduate student, the graduate student, and the interested professional plant physiologist an introduction to recent studies in which the techniques of molecular biology have been applied to classical problems of plant development. We have not seen another book like this one, and it is easy to understand why. Up to five or six years ago, there was not enough information. Molecular biology was clearly an exciting field, but few laboratories had performed and published results of experiments with plant tissues. Today the field is moving so quickly that almost every discovery is outdated by the time it appears in print, a frustrating situation for an author.

We envision this book being used in one of three ways: 1) as a supplement to a plant physiology class, providing a new perspective toward growth and development; 2) as a supplement to a class in developmental biology, providing information about plants to an audience that otherwise would learn only about bacterial and animal systems; 3) as a basic book for a specialized course in plant molecular biology, perhaps to be supplemented with more recent research reports and proceedings from workshops. We hope too that researchers in various fields may find it to be a useful reference.

The first two chapters are introductory, providing a botanical background for the developmental biologist and background in molecular genetics for the classically-trained botanist or plant physiologist. The third chapter describes some selected examples of plant development, which are analyzed in terms of induced gene expression. The fourth and fifth chapters present the overall organization of plants' nuclear and organellar genomes, along with a detailed description of certain genes and gene families. More advanced information about the control of gene expression, as inferred from gene structure, is found in these chapters. The sixth chapter describes some recent work on genetic engineering, not because engineering is the subject of this book, but because the techniques of transformation provide a new tool for analyzing the control of development.

We owe much to the many plant molecular biologists on whose work this book is based. We cannot name them all, but we can extend a special appreciation to those who sent us photographs or permitted us to use figures from their publications: Lisa Baird, John Bedbrook, Chris Bornman, Anthony Cashmore, Richard Flavell, Michael Freeling, Wayne L. Gerlach, Robert Goldberg, Milton Gordon, Wilhelm Gruissem, Klaus Hahlbrock, Barbara Hamkalo, John Harada, Tuan-Hua David Ho, William J. Hurkman, Lon Kaufman, Hans Kössel, Anton Lang, Brian Larkins, Christopher J. Leaver, Arieh Levy, Danny Llewellyn, April McCoy, Joachim Messing, Oscar Miller, Jr., Judy Murphy, John Oross, Jeffrey Palmer, James Peacock, Charles Rick, Ferrucio Ritossa, Robert Shepherd, Charles Sloger, George R. Stark, David Stern, Desh Pal Verma, Geoffrey Wahl, Andrew Wang, Barbara Webster, Jon Watson, John Yoder, and J. Zischke.

We would also like to thank the following reviewers for their helpful comments: Wilhelm Gruissem, University of California, Berkeley; Dawn S. Luthe, Mississippi State University; and Steven Spiker, North Carolina State University.

# Molecular
# Plant
# Development

# ONE ———————————

# Developmental Stages
# in the Plant Life Cycle

## A. THE LIFE CYCLE OF HIGHER PLANTS

All living organisms face certain basic problems. Living organisms must get energy for powering the chemical and physical processes of life. They must find chemical elements for synthesizing organic materials. And they must deal with stressful aspects of the environment, such as excessively high or low concentrations of water. In contrast to animals, which obtain their energy and many of their chemicals by eating organic matter (both dead matter and other living organisms), plants are able to satisfy their needs for energy and chemical mass entirely from the inorganic environment. The plant lifestyle is, therefore, said to be *autotrophic* (self-feeding).

Many of the details of plants' life cycles, including mechanisms of growth, development, and reproduction, can be understood as adaptations of an autotrophic lifestyle. For instance, plants obtain energy and carbon through the process of photosynthesis. Exposing photosynthetic tissues to light and obtaining an adequate supply of carbon dioxide involves the development of a shoot system. For land plants, it is usually impossible to absorb carbon dioxide without also losing needed water. Plants can get more water (as well as mineral nutrients) from the soil through the development of a root system. A root system fixes a plant in a specific place, which causes the development of special mechanisms for dispersing their progeny, such as flowers and fruits.

Plant cells are often bathed in water that is purer (water at a higher concentration and with fewer solutes) than the water within their cytoplasm. As a result of this situation, osmosis tends to force a net flow of water into the cell across the plasma membrane. If this tendency were not opposed, the cell would swell and explode. But the plant cell wall provides the mechanical force that opposes the inward water flow. The plant cell wall represents an adaptation that significantly influences a plant's lifestyle. Plant cell walls limit motility, so most plants move and react more slowly to their environment than do animals. Plant cells are fixed in place within their tissues by their cell walls, so that strategies of growth and development that involve movement of cells throughout the body, so common in animals, are not found in plants.

The life cycle of flowering plants includes a number of discrete stages arranged in sequence, such that one stage leads directly, though not necessarily immediately, to the next stage. Since a life cycle *is* a cycle, there is no obvious starting or ending point, as is clear from the diagram in Figure 1–1. However, since most of us think of growing plants from seed, the seed stage is a reasonable point to begin our discussion of the developmental processes that occur in each stage of the life cycle.

The seed contains a small embryonic plantlet in a state of suspended animation. The plantlet includes tissues destined to become the shoot system and tissues destined to become the root system. In addition, there are one or two cotyledons, which contain some storage compounds that nourish the embryo when it starts to grow. In some plants the cotyledons also become functional leaves. Shoot, root, and cotyledon(s) all contain diploid cells formed by mitotic cell division from a single zygote cell. In addition, there are also other tissues in the seed that originate from different sources. The endosperm is another storage tissue. It is formed from a separate fertilization event, which joins one nucleus from the pollen and two nuclei related to the egg nucleus but separate from it (discussion follows).

The relative importance of endosperm and cotyledons as storage tissue varies among species of plants. In monocots (plants with one cotyledon in their seed, for example, grasses) the endosperm is usually the major storage tissue, whereas in dicots (plants with two cotyledons per seed, for example, most broadleaf plants) the cotyledons usually perform this function.

The seed is covered with a seed coat, which is formed from maternal tissues of the mother plant. In general, the seed is very dry, usually containing less than 50% water. The seed is dormant, though some degree of metabolic activity can be detected.

The seed germinates in response to an environmental signal. Often this signal is simply the presence of water, but sometimes other factors are important, such as illumination, a period of cold, or a breaking of the seed coat.

The germination of a seed is complex. It involves the "mobilization" of storage compounds, that is, their biochemical breakdown into components that can be transferred to the shoot and root of the embryo. Often the enzymes needed for the breakdown are not present and must be synthesized. Storage proteins will be metabolized into amino acids; starch will be hydrolyzed into glucose; oils will be oxidized and converted into acetyl CoA, and the acetyl CoA will be converted into sugars.

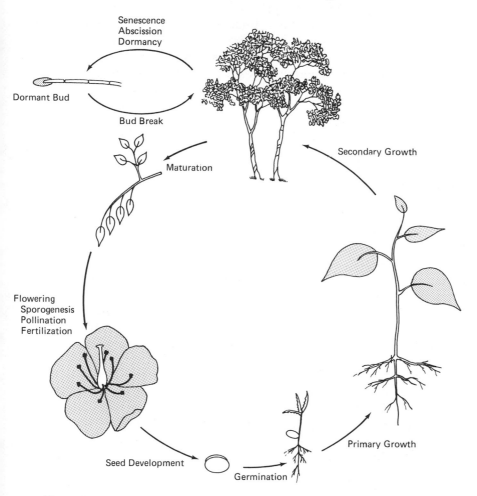

**Figure 1-1.** Life cycle of a flowering plant. Bottom: A seed germinates in the appropriate environment, and the shoot and root of the seedling grow in length through "primary growth" processes (see Fig. 1–2). "Secondary growth" allows thickening of stems and roots. Formation of a "dormant bud" allows perennial plants to live through harsh conditions. Shoots must undergo "maturation" before they can flower. The sexual processes that occur in the flower (see Fig. 1–3) produce an embryo that, together with maternal tissues, develops into a seed.

Coordinated with the mobilization of the storage compounds, there is a rapid initiation of cell division in the shoot and root tips of the embryo. There is elongation of the new cells. The pressure that drives the elongation of the cells pushes them through the seed coat and out into the environment.

If the embryonic tissues emerge from the seed into a dark world, as may happen if the seed is deeply buried, they become etiolated and are significantly different from normal tissues in the light. This is particularly true for shoot tissues. Etiolated shoots are pale yellow instead of green, and they have partly developed plastids (etioplasts). They also lack or have low amounts of the enzymes required for photosynthesis. Dicots typically have long stems but very small leaves. Monocots have elongated and thin leaves that are covered by a sheath, called the "coleoptile." The rapid elongation of the shoot allows it to grow out of the soil and into the light quickly. When the shoot tissues reach the light, there is a rapid change in their developmental pattern. Stem elongation slows; leaf cells expand. Etioplasts develop thylakoid membranes and all the membrane proteins and soluble enzymes needed for photosynthesis.

The growth of roots and shoots involves three processes, which can be separated logically and, to some extent, physically from one another (Fig. 1–2). These processes are cell division, cell enlargement, and cell differentiation. In roots and in many shoots, cell division occurs primarily at the apex in a region called the apical meristem. Cells produced by the apical meristem move away from the meristem; or more correctly, the meristem moves away from them. In a region directly behind the apical meristem, these cells begin to enlarge. During and after enlargement, the cells differentiate into specialized cells, such as xylem tracheids, sieve elements of the phloem, trichomes on the epidermis, and photosynthetically competent parenchymal cells. The regions in which cell division, elongation, and differentiation take place and the degree to which they can be separated vary among different plants. In grasses, for instance, there is a meristem at the base of each leaf. Cells produced at the base elongate and differentiate as they are thrust up away from the base.

The primary growth of young roots and shoots is called "indeterminate," because it has no clear limit. In a dicot shoot, cell division, enlargement, and differentiation can continue to produce stem and leaves so long as the shoot is well nourished and no environmental signal changes the pattern of growth in the meristem. This is in contrast to the growth of animals. Animal species are limited in the size the adults will reach, in the number of organs that will develop, and in the time period over which the individuals can grow.

In the primary roots and shoots, there are three main tissue regions: epidermis, ground tissue (called cortex and pith), and vascular bundles (called the stele in roots). In each tissue there are several cell types, which are formed by differentiation from the cells produced by the meristem. The shoot epidermis has epidermal cells specialized for protection (production of cuticle), guard cells, and trichomes (hair cells). The ground tissue includes parenchymal cells and some cells for support, such as the collenchyma. The vascular bundles contain tracheids and fibers, vessel elements, sieve elements with their companion cells, and parenchymal cells. There are also specialized organs associated with primary growth, for instance, the root cap. Root cap cells are produced by the meristem and move forward along the direction of growth. These cells produce and secrete slime (a polysaccharide), and then die and fall off the root.

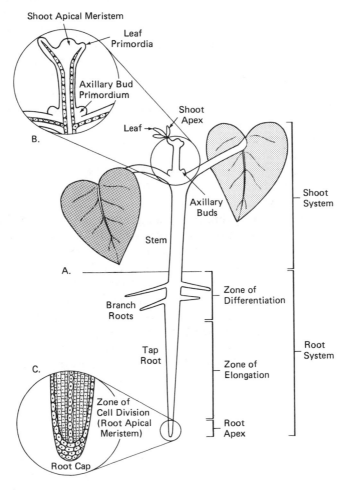

**Figure 1-2.**   Primary growth of a dicot plant. A. Overall morphology of a small bean seedling. B. Longitudinal section of the shoot apex. C.  Longitudinal section of the root apex.

Complex shoot and root systems can be formed by branching. This involves the formation of a new apical meristem or activation of a dormant one. In roots the pericycle, a layer of cells surrounding the vascular system, can form the new meristem. The cells that will form a new meristem begin active cell division in a small area and push themselves through the original cortex and epidermis. These cells that they form enlarge and differentiate like the original root, to which of course they are connected. In shoots, there are specialized organs that are potential apical meristems for branches. These are the axillary buds, located in the "axils," the upper angles formed where the leaf petioles connect to the stem. Axillary buds are dormant when formed, but when they are activated they start to produce a shoot system just like that formed by the original apex. Part of the activation process involves the development of vascular connections between the main stem and the branch. Branches can also form in unusual places under unusual circumstances and normally in particular plants. Adventitious roots can appear from a shoot that has been cut from its root system or in some cases from shoots that are close to the ground. Adventitious shoots appear from some underground roots. In these cases, the adventitious organ represents the reorganization of mature, differentiated tissue into a new meristem.

In dicots, after primary growth, there is secondary growth and development that allows for the radial thickening of both shoots and roots. There are two new meristems involved in producing the new cells. The first, the vascular cambium, is a cylinder of cells that runs between the xylem and phloem of the primary vascular bundles. As cambial cells divide, they produce non-dividing cells both to the inside and to the outside of their cylinder. The cells on the inside develop into xylem cells, most of which are tracheids and vessel elements. The cells on the outside develop into phloem cells, including sieve elements, companion cells, and some parenchyma. The second new meristem, the cork cambium, forms from the epidermis or the outer cells of the cortex. These cells form a rough cylinder outside the vascular cambium cylinder. Cork cells produced to the outside take the place of the primary epidermis. They accumulate a hard, waxy substance, called suberin. After these cells die, they leave the suberin as a protective coating.

Many plants will alternate periods of vegetative growth, both primary and secondary, with periods of dormancy. Generally, dormancy occurs in winter, when light and temperature are not conducive to growth. In preparation for dormancy, the apical meristem stops its normal vegetative growth processes and forms a special organ, the dormant (or resting) bud. The dormant bud has several protective scales, analogous to leaves and formed in a similar way, which cover a live, but inactive meristem. In spring, when the cold weather has passed and there is more light, cell division in the meristem may resume. This leads to normal primary shoot growth.

There are other events associated with dormancy, including the senescence and abscission of leaves. In preparation for dormancy, many materials in leaves, such as chlorophyll and proteins, are mobilized—that is, broken down and transported to storage areas in stems or roots. The loss of chlorophyll leads to the spectacular color change of leaves in the fall. Abscission (leaf fall) occurs after the mobilization of the storage materials. A specialized layer of cells in the leaf petiole produces hydrolytic enzymes that destroy cell walls and weaken the supporting structure of the petiole, and the leaf easily falls off the stem. In the spring, new leaves must be formed from new primary stem growth.

The next stage in the life cycle is the induction of flower development. This involves a change in the pattern of development of an apical meristem, either one located at the tip of the main shoot or one at an axillary position. The triggering of the change remains one of the mysteries of plant development. We know that it requires that the shoot be mature. In some plants, like ivy, mature shoots can easily be distinguished from juvenile shoots by the shapes of their leaves. In some plants maturation requires a treatment of the apex with cold temperatures. And in some plants it requires a change in the length of days and nights. The days must become long enough for spring flowering plants or short enough for fall flowering plants.

Once a plant has been triggered to flower, the apical meristems that are involved stop their indeterminate growth pattern and adopt a determinate one. They cease producing leaves and instead produce a well-defined series of organs that make up the flower. These organs include the sepals (collectively, called the calyx), petals (the corolla), the stamens (the androecium), and the carpels (the gynoecium). Developmentally, each set of organs (calyx, corolla, androecium, gynoecium) is considered analogous to a whorl of leaves, but the organs are quite different from leaves and from each other in number, size, shape of cells, and in the patterns of differentiation they exhibit.

Sexual events of the plant life cycle occur in the tissues of the carpels and stamens (Fig. 1–3). Within the carpel a mass of cells, the ovule, forms. And within the ovule, one cell, the megasporocyte, undergoes meiotic division. The haploid products of meiosis in plants are "spores"; the products of meiosis of the megasporocyte are megaspores. One megaspore enlarges and produces eight nuclei through three sequential mitotic divisions. The collection of eight haploid nuclei (and their attendant organelles) make up the "megagametophyte," which is equivalent to the larger multicellular haploid stages found in the life cycles of more primitive plants. The megagametophyte produces the female gamete; that is, one of its nuclei becomes the egg cell. Within the anther, the organ at the tip of the stamen, there are many cells, microsporocytes, that undergo meiosis. The products of meiosis, microspores, split apart and develop the complex cell walls characteristic of pollen grains. The nucleus within each

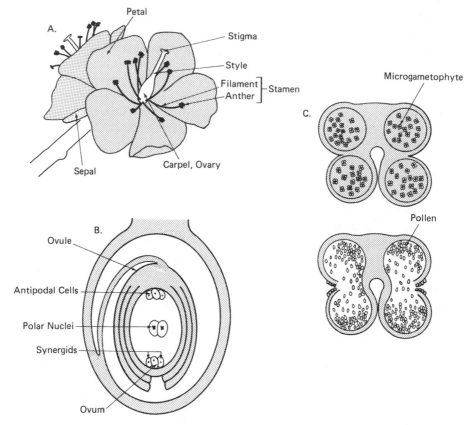

**Figure 1-3.** Sexual development within a flower. A. Parts of the flower. B. Section of carpel enlarged to show the macrosporophyte and its eight haploid nuclei that are formed by meiosis (plus mitosis) of a single macrosporocyte. C. Section of anther showing four haploid products of meiosis of a microsporocyte and pollen grains formed from each of the haploid meiotic products.

pollen grain divides by mitosis once or twice, forming a haploid microgametophyte, and one of the resulting nuclei becomes the generative (sperm) nucleus. Note that this process differs from gametogenesis in animals. In animals, gametes are formed directly from the products of meiosis: There is no multicellular haploid stage.

The next step in the plant's life cycle is fertilization, the union of egg cell with sperm nucleus. This may be a complex process. Movement of a pollen grain to the stigma at the tip of a carpel generally requires the movement of an outside agent, perhaps the wind or an animal. Once the pollen grain arrives, there may be a selective process to insure that it can complete fertilization only if it is of the correct species. The selection may involve an interplay between proteins on the surface of the stigma with those in the cell wall of the pollen.

If the pollen and stigma are compatible, the pollen germinates and forms a tube, which elongates and pushes through the tissues of the carpel until it reaches the ovule and egg cell. The tube penetrates the egg cell, releasing the sperm nucleus, which then fuses with the egg nucleus. A second haploid sperm nucleus fuses with two haploid nuclei (polar nuclei) of the megagametophyte, forming the endosperm.

Following fertilization, there is rapid cell division by the zygote: This forms the embryo, including the root tissues, shoot tissues, and the cotyledons. There is also rapid division of the endosperm nuclei. During the latter part of this period of cell division and after it, there is a period in which the endosperm or the cotyledons grow rapidly as they store nutrients (proteins, starches, and fats) which they make from materials supplied by the mother plant. Then the surrounding maternal tissues, the integuments, harden to form the seed coat. These tissues then lose water, and the seed is mature.

Following fertilization, the ovary (the tissue of the gynoecium that surrounds the developing seeds) also develops rapidly, forming a fruit. The purpose of the fruit is to protect the seed and to aid in its dispersal, for instance, by attracting animals that eat the fruit and carry the seed to a new location. The processes of fruit development vary widely among plants, but they often include a burst of cell division and cell enlargement, followed by the senescence of the new cells. They may include the synthesis of chemical compounds that alter the color, taste, or smell of the fruit.

The life cycle of a higher plant shows that there are many different aspects to plant development. Within each stage of development there are many different tissues, and within each tissue there are several types of cells. We must learn how cells and tissues form within a given organ. Sometimes through experimental manipulations we can study development at the cell level, isolating cells and studying the control of mitotic division, expansion, and differentiation. Sometimes we must study development at the organ level, analyzing the relationships among different types of cells to learn the factors that control development in the natural situation. Some developmental events in plant life cycles are summarized in Table 1–1.

Another point to study is the progression from stage to stage within the life cycle. What controls seed germination, or leaf senescence, or flowering? Sometimes the triggers are internal, controlled by time alone, independent of environment; but often they are related to the environment. Much has been learned about the environmental factors that produce a given response, but little is known about how these factors actually influence basic processes like cell division, elongation, and differentiation.

TABLE 1-1  **Some Developmental Events in Plant Life Cycles.**

| Stage | Tissue Location | Event |
|---|---|---|
| Seed germination | Endosperm, cotyledons | Mobilization of fats, starches, proteins |
| | Embryo | Cell division |
| | | Cell expansion |
| Primary growth and development | Meristem | Cell division |
| | Below apical meristem | Cell elongation |
| | Leaf | Cutin synthesis |
| | Stem epidermis | Differentiation of guard cells |
| | | Differentiation of secretory glands |
| | Leaf mesophyll, stem cortex | Chloroplast differentiation |
| | Vascular bundles | Differentiation of fibers, tracheids, vessels |
| | | Differentiation of sieve tube elements |
| Secondary growth and development | Vascular cambium | Cell division |
| | | Differentiation of fibers, tracheids, vessels |
| | | Differentiation of sieve tube elements |
| | Cork cambium | Cell division |
| | | Suberin synthesis |
| Dormancy | Apical meristem | Dormant bud formation |
| | Leaves | Senescence |
| | Petioles | Abscission |
| | Roots | Storage of starch and protein |
| Flowering and fruit development | Corolla | Synthesis of flavonoids |
| | Ovule, anther | Sporogenesis |
| | Ovary | Cell division |
| | | Cell expansion |
| | | Synthesis of colored and volatile compounds |
| | | Cell wall digestion |
| Seed development | Endosperm, cotyledons | Synthesis of fats, starches, proteins |

## B. CLASSICAL CONCEPTS OF DEVELOPMENT

When a multicellular organism develops from a single cell, either spore or zygote, the single cell is said to be *totipotent*—it contains all the information necessary to make the entire organism. In sea urchins, the cells formed from the first two or three cleavages of the zygote are also totipotent. If these cells are separated, they can each continue to divide, and eventually each will form a whole adult organism. After several divisions, however, animal cells lose totipotency. If animal cells are separated from their tissue they may proliferate, but they will not form all the cells and tissues needed to make another organism. Moreover, some animal cells, for example neural or epidermal cells from a late-stage gastrula, when moved from one place on an embryo to another, continue to form the organ they would have formed in their original position. These cells, and their developmental fates, are said to be determined. The determined cell is stable, and its characteristics are not altered by changing its environment.

The concept of determination arose at the beginning of the twentieth century when A. Weismann proposed that determinants (genes) in the germ plasm of a zygote were distributed among the products of the cleavage divisions. Demonstrations of totipotency disproved this mechanism, but the concept persisted. Ian Sussex has defined determination as an event that (a) occurs at high frequency, (b) is directed, (c) is relatively stable, and (d) is transmitted clonally, that is, by mitosis. Whether the concept of determination applies to plants as it does to animals is a matter that will be considered subsequently.

The concept of totipotency is easy to apply to higher plants. Not only is the zygote totipotent, but many multicellular pieces of tissue from a growing plant—and even single cells—are totipotent also. Indications that shoots and roots are not irreversibly determined have been recognized for a long time. Cuttings of shoots of many plants will develop roots at their basal ends. This is the basis of the commercial propagation of roses. In addition, roots severed from many weedy species will sprout new shoots. It is clear that shoots and roots have cells that maintain the capability of redifferentiating into other organs. However, it is not clear from these observations whether this capability is vested in a single cell or whether it depends on a preexisting tissue organization.

Further evidence on totipotency has come from the culture of plant cells and plant tissue on artificial medium, where smaller pieces of tissue and single cells can form whole plants. In the 1950s, F. Skoog, F.C. Steward, and others defined the nutrients necessary for the growth of plant tissues. These nutrients include inorganic salts, sugar as a carbon source, certain vitamins like thiamine and inositol, and the growth regulators (hormones) auxin and cytokinin. If a single meristem is divided into six parts and each part is placed on the appropriate medium, each will form a whole meristem and begin growing. In medium containing the appropriate concentrations of these nutrients, the explants of shoots or roots will form callus, a mass of relatively undifferentiated cells. The callus often is friable and can be broken into small pieces that can be placed on more medium for further growth. Skoog showed that a change in the relative concentration of the growth regulators caused tobacco callus to reform shoots *or* roots, depending on the change. Steward demonstrated that a suspension of carrot cells, actually small pieces of callus in a liquid medium, could form embryoids when the growth regulators or nitrogen sources were changed. It was clear, particularly from Steward's work with carrots, that very many of the cultured cells were totipotent, possessing the ability to form whole plants.

Since virtually all plant cultures have cells in clumps, the early culture studies did not completely resolve the question of whether totipotency resided in single cells or required the participation of several cells. Steward eventually demonstrated that a single carrot cell could form an embryoid. More recently, it has been possible with several species to isolate single cells from suspension cultures, or from leaves, by using hydrolytic enzymes that break cells apart by cleaving cell wall components. Some of these cells can reform walls, divide to form callus, and eventually develop shoots and roots. This is most successful with species of the Solanaceae family, such as tobacco, petunia, and potato, but it has been performed with members of other families as well (Fig. 1–4).

Thus in a number of cases, plant cells seem to be totipotent and indeterminate. This has led certain authors to suggest that the concept of determination—to the degree that it suggests a loss of developmental potential—may not apply to plants. However, there are cases where one might wish to apply the concept of determination.

First, it should be noted that the number of culture systems that can be induced to regenerate whole plants is at present limited. Even among those shown to regenerate, there are very few in which the regenerant comes from a single, defined cell. Furthermore, it is a common observation that plant tissues kept in culture for an extended period (from weeks to years, depending on the plant) lose the ability to regenerate whole plants. This is sometimes, but not always, associated with changes in the chromosomal complement or structure of cells that are kept in the culture. The balance of evidence suggests that not all cells in culture are totipotent.

Positive evidence of determination may be found in the structure and behavior of cells in an intact plant. There are certain cells which clearly lose totipotency when they develop. These include xylem elements, which die in the final stages of their development, and the sieve elements of phloem, which lose their nuclei. These do not fit the definition of determined cells mentioned above, because they cannot be clonally propagated. But other cells, such as hair cells on shoot epidermis, are difficult to culture and thus they may also be determined. The existence of totipotent parenchymal cells in plants does not preclude the possibility that other cell types may be determined in the sense of having restricted developmental potentiality.

There is a different way of looking at determination, and one that applies well to plants. One can see a determined cell as one that has acquired the ability to develop along a given path, often in response to a specific induction signal. In animal embryos, the presence of cells from

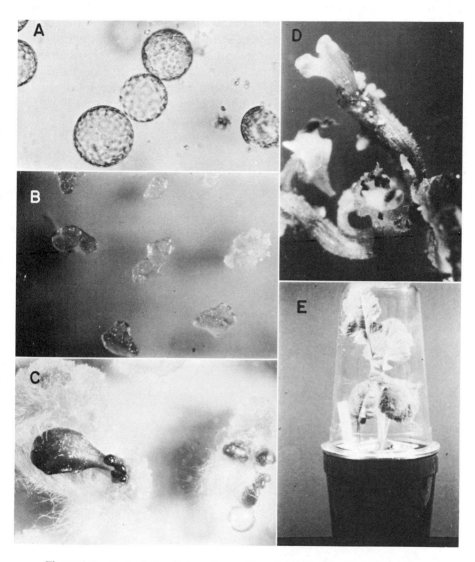

**Figure 1-4.**    Protoplasts of oil seed rape (*Brassica napus* L.) and plants regenerated from them. A. Isolated protoplasts from petiole mesophyll. B. Callus (5–6 weeks old) on Murashige-Skoog medium with 0.5 mg/l naphthalene acetic acid as auxin and 1.0 mg/l 2-isopentenyl adenine as cytokinin. C.   Adventitious shoots and roots (6–10 weeks) on Murashige-Skoog medium with hormones. D. Plantlets (10–12 weeks). E. Plant in pot under hardening conditions. Photographs kindly provided by Dr. Chris Bornman.

one region induces development in another. For instance, cells from the dorsal lip of the blastocoel induce the ectoderm to form neural tissue. The developmental pathways available to the ectodermal cells are pre-determined; the specific pathway is induced by chemicals from the dorsal lip cells. In plants, the signal for induction may be light or temperature or the presence of a specific hormone. Plants have at least five classes of hormones, including the auxin and cytokinin mentioned previously in the discussion on culture media. A common observation is that cells in different tissues react differently when exposed to a given hormone, suggesting

**Figure 1-5.** The effect of ethylene on different tissues. Top: ripe and unripe figs. Fig ripening is stimulated by ethylene, either from the plant or added artificially. Bottom: epinasty in tomato plants. Compare the petioles of the ethylene-treated plant (left) to those of the control (right).

that each cell type has a determined, specific reaction pattern that can be induced by the hormone.

A good example involves the hormone ethylene (Fig. 1–5). Ethylene applied to a full-sized, but unripe tomato fruit will cause it to ripen. The ripening response involves the softening of cell walls, the loss of chlorophyll, the production of xanthophylls, and the conversion of starch to sugar. The cells of the mature fruit are said to be "competent" to respond to ethylene. Ethylene applied to an immature tomato fruit will not give the same response—the cells are not able to respond. Ethylene applied to the shoot system of a tomato plant will cause a very different response: A selective cell enlargement in the leaf petioles will cause the leaves to bend downward (epinasty). Further, if the plant were under some nitrogen stress, the mature leaves might have taken a step toward senescence. In this case, the ethylene might trigger a special layer of cells in the petiole to autolyze, releasing quantities of hydrolytic enzymes that destroy the petiole cells' walls and allow the leaves to fall from the stem. It is clear that much of the specificity of plant development depends not on the type of hormone involved but on some previous determination of a cell's biochemical and physiological response once it contacts the hormone.

Thus far, the concepts of totipotency, determination, competency, and induction have been defined and discussed in rather vague and very empirical terms. Do these terms define objective states of plant cells? If so, it should at some point be possible to distinguish a totipotent from a determined cell in terms of its structure or biochemistry and not just in terms of its behavior.

## C. REGULATORS OF GROWTH AND DEVELOPMENT

Hormones have a special place in the subject of plant development. Much of the work on plant development by physiologists has concerned the nature, localization, and effects of these growth regulators. There are major questions about the roles of hormones that remain unanswered. One question we asked in the previous section: How is it that different cells respond in different ways to the same hormone? We might also ask why a particular developmental step, such as flowering, is triggered by different hormones in different species of plants. The following discussion provides some background on these perplexing observations.

Auxin, the first plant hormone to be discovered, is generally identified as a growth-stimulating substance. When sections of young plant organs such as oat coleoptiles, pea stems, or soybean hypocotyls are treated with auxin, they rapidly start to lengthen. This is caused by the elongation of each individual cell (Fig. 1–6). In these organs, the enlargement of the cells is polarized—confined to one dimension—by the structure of the cells' walls. Higher concentrations of auxin applied to stems or

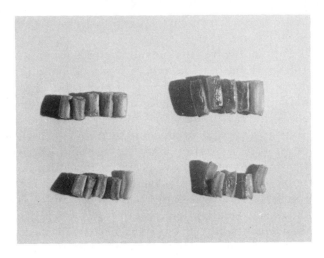

**Figure 1-6.** Effect of auxin on pea stem elongation. Top: third internode. Bottom: fourth internode. Left column: control without auxin. Right column: treatment with 0.01 mM indole acetic acid.

leaves, especially in conjunction with a wound, stimulate isodiametric (spherical) cell expansion and cell divisions to form an abnormal tissue mass or callus.

A natural auxin, indole acetic acid (shown as IAA in Fig. 1–7), is synthesized enzymatically from the amino acid tryptophan. While all plant cells have the genetic ability to produce IAA, only a few show this ability. In vegetatively growing plants, most auxin is produced by young leaves in the shoot apices. Cells in older leaves, stems, and roots do not make detectable amounts of auxin. Auxin is also synthesized by developing embryos after pollination and fertilization.

The concentration of auxin in other parts of a plant depends on their proximity to the sources of auxin production, on the presence and activity of auxin-inactivating enzymes, and on the activity of a very specific transport system. Thus, auxin moves basipetally from the shoot tips to promote elongation of the stem and root. And auxin also moves out from developing embryos to promote fruit growth. Auxin carriers in the membranes of certain parenchymal cells mediate the flow of auxin from cell to cell. The activities of these carriers can often be influenced by gravity and light. In at least some cases this is the basis of gravitropic and phototropic growth. For instance, a beam of blue light directed onto one side of an oat coleoptile stimulates transport of auxin toward the dark side away from the light. The cells on the dark side expand more rapidly than those on the illuminated side, and as a result the coleoptile bends toward the light.

Though cell expansion is the most characteristic effect of auxin, this hormone has other effects on plant development. These effects can be seen by removing the source of auxin (usually the shoot apex) and replac-

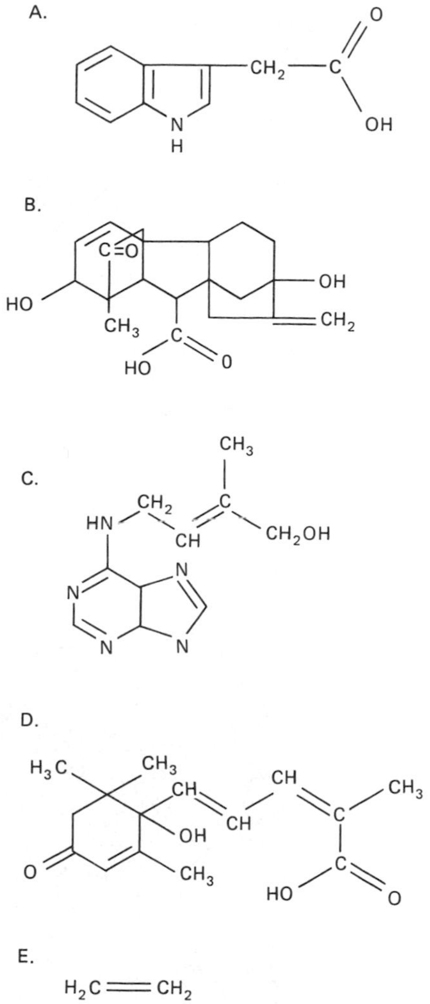

**Figure 1-7.** Natural plant hormones. A. Indole acetic acid (auxin). B. Gibberellic acid, GA$_3$. C. Zeatin (cytokinin). D. Abscisic acid. E. Ethylene.

ing it with an artificial auxin source (such as IAA in an agar gel). Such experiments suggest, for instance, that auxin inhibits the growth of axillary shoot buds. Normally bud growth starts when the shoot apex is removed, but if the apex is replaced by an artificial auxin source the bud growth stops. Similar experiments indicate that auxin promotes the initiation of branch root primordia, stimulates cell division in the vascular cambium, and inhibits the abscission of leaves.

How auxin affects different parts of a plant in such different ways remains a mystery. In some cases, such as axillary-bud inhibition, the effect of auxin may be indirect, the result of growth stimulation in another part of the plant. However, the possibility of different direct effects on different cell types cannot be ruled out.

More discussion of the mechanism of auxin action can be found in Chapter 3, Section D.

Gibberellins are another class of hormones with generally growth-promoting properties. Several lines of evidence indicate that gibberellins are required for normal shoot elongation. Certain plants, such as spinach and cabbage, have a "rosette" morphology in which the central stem is very short. Even though many leaves have formed, the internodal regions of the stem have not elongated. The plants keep this form until just before flowering, when the stem "bolts," or rapidly lengthens. Spraying a solution of gibberellin on such plants produces a rapid lengthening of the stem, much like the bolting that occurs before flowering (Fig. 1–8). There are also certain mutants of corn and pea that have short internodes, and spraying these plants with gibberellin stimulates their stems' growth, too. Conversely, applying chemicals that block the synthesis of gibberellin stunts the stem growth of non-rosette plants, like chrysanthemum. It appears that all plants need gibberellins for shoot elongation. Many produce it constantly during their growing period, but some rosette plants control their morphology by regulating the timing of gibberellin synthesis.

**Figure 1-8.**   Effect of gibberellin on stem growth and flowering in carrot (*Daucus carota* L.). Left: control. Middle: 50 mg/l GA with no cold treatment. Right: cold treatment. Photograph kindly provided by Dr. Anton Lang. (See *Proc. Natl. Acad. Sci.* USA 43:709, 1957.)

Gibberellins are lipid-soluble, multi-ring compounds (Fig. 1–7) formed from acetyl CoA through the mevalonic acid pathway. Like auxins, gibberellins are produced only in certain tissues, and these are generally the same tissues that make auxin: young leaves and developing embryos. Gibberellins are transported throughout the plant, probably through the phloem.

Gibberellins, like auxins, have a number of effects on plant development. They stimulate flowering in certain rosette plants, though this may be an indirect consequence of the bolting they induce. They also stimulate spring bud break in many dormant species. Like auxins, they stimulate cell division in the vascular cambium.

Gibberellins produce a striking developmental effect in the germinating seeds of barley and other members of the grass family. Much of the nutrient of these seeds exists as starch in the endosperm. This starch must be metabolized to smaller sugar molecules that can be used by the growing embryo. The metabolism of the starch is mediated by gibberellins. Soon after the seed is hydrated, the embryo produces and releases gibberellins, which diffuse to the aleurone, or outer, layer of the endosperm. In the presence of gibberellin, the aleurone layer synthesizes and releases alpha-amylase. The alpha-amylase, together with other enzymes, breaks down the starch in the inner endosperm tissue. Thus, gibberellins help to coordinate developmental events in different tissues of the seed. Isolated aleurone layers are also a favorite material for studying the mechanisms of gibberellin action.

Cytokinins are a third class of growth stimulators. These compounds were discovered because of their ability to promote cell division in cultured plant callus tissues. This activity explains why they are normally included in plant tissue culture media, as was mentioned in the previous section.

The synthesis of cytokinins involves a modification of the nucleic-acid base, adenine, by the addition of an isopentenyl group (from the mevalonic acid pathway) to the N attached to the 6–carbon (Fig. 1–7). Like other hormones, cytokinins are produced only in certain tissues. Most are made in the roots and transported to the shoots in the transpiration stream. They are also synthesized by developing embryos, and endosperm is a particularly rich source.

Cytokinins are thought to be required for cell division in the various meristematic regions of the plant. A solution of cytokinin applied to an inhibited axillary bud of a pea plant can produce a transient stimulation of the bud's growth. This activity may be related to the hormone's stimulation of cell division. Cytokinins have another property that has no obvious relation to cell division: the inhibition of senescence of green tissue. Leaves that are detached from their plant and kept moist normally turn yellow in a few days, but a spray of cytokinin on the leaf markedly delays the yellowing. This effect involves a slowing down of the synthesis of proteases and chlorophyllase that normally occurs in the detached leaves, but it occurs together with a stimulation of the general protein-synthetic capacity of the tissue. Cytokinins also have the ability to stimulate the enlargement of radish cotyledons. This seems to involve the enlargement of individual cells, so cytokinin seems to act on these organs in the same way that auxin acts on oat coleoptiles, but the relationship between cytokinin- and auxin-stimulated growth is not certain.

Abscisic acid (ABA in Fig. 1–7), a fourth hormone, is produced in leaves from compounds synthesized in the mevalonic acid pathway.

ABA was named for its ability to promote the abscission of cotyledonary petioles from the stems of cotton seedlings. It promotes abscission of leaves in some other plants (Fig. 1–9), and it has other activities as well. A major developmental function is the induction of a resting bud and the promotion of dormancy in perennial species at the end of their growing season. The timing of dormancy is related to the fact that ABA accumulates in leaves and is transported to the shoot apex as days grow shorter in fall. Another function is the maintenance of dormancy in certain seeds. Such seeds cannot germinate until the ABA has leached away or been metabolized. Often a long cold treatment under moist conditions is needed

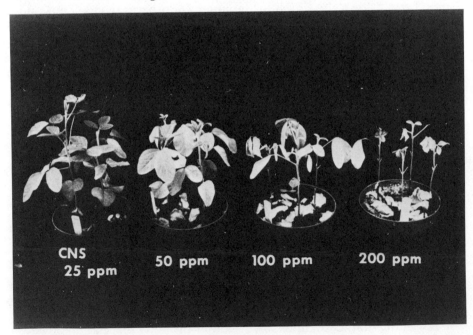

**Figure 1-9.** Effect of abscisic acid on leaf abscission in soybean (cultivar CNS) plants. Plants were sprayed with ABA solutions at the indicated concentrations. Reprinted from C. Sloger and B.E. Caldwell, *Plant Physiology* 46:634, 1970.

to induce the metabolic disappearance of this hormone. Thus, ABA has several functions generally related to the inhibition of growth under certain environmental conditions.

ABA has another function that seems quite unrelated to the long-term developmental effects mentioned previously. It causes a rapid closing of leaf stomata when a leaf begins to wilt. It appears that a turgor-pressure sensor in the leaf cells somehow controls the activity of ABA-synthetic enzymes. At the first sign of wilt (low turgor pressure), ABA synthesis accelerates, and within minutes the concentration of the hormone becomes very high. The ABA acts directly on the guard cells, causing them to lose $K^+$ and other solutes. The consequent loss of turgor pressure in the guard cells, relative to the turgor pressure in surrounding cells, allows each pair of guard cells to move together and close off their stoma. The high concentration of ABA produced during wilt generally does not lead to long-term developmental effects, because rehydration of the leaves stimulates enzymes that break down the hormone.

In many cases, ABA seems to antagonize the effects of gibberellins. This can be seen in dormant perennial plants, in which spring bud break is associated with both a loss of ABA and an increase of gibberellins. A more direct interaction is seen in barley aleurone layers, where ABA slows the induction of alpha-amylase by gibberellins.

Another major plant hormone, ethylene, has a number of activities. Several of these were mentioned in the previous section: fruit ripening, epinasty, stimulation of abscission. Fruit ripening is one specialized example of senescence, and ethylene can also stimulate senescence in other tissues, like detached green leaves and flower petals. Because it stimulates senescence and abscission, ethylene is generally considered a mediator of growth inhibition. However, it has one effect that does not fit this pattern: stimulation of flowering in bromeliads, like pineapple.

On the cell level, ethylene stimulates isodiametric enlargement of cells by interfering with the polarized deposition of cell wall microfibrils. This leads to short, fat stems and roots under conditions where ethylene is synthesized rapidly or otherwise accumulated in high concentrations. Ethylene inhibits auxin transport, too, so the fat stems and roots do not show normal geotropic and phototropic curvature.

Ethylene is formed from the amino acid methionine through a biochemical pathway that involves the intermediates S-adenosyl methionine and aminocyclopropylcarboxylic acid (ACC). Most, if not all, plant tissues have the ability to synthesize ethylene, but they show the ability only at particular times or under stressful conditions. Wounding stimulates ethylene synthesis, and so does high auxin concentration, explaining why callus with isodiametric cells forms around wounds and after treatment with high auxin concentration. Lack of oxygen stimulates ethylene synthesis in tomato roots, and tomato plants with waterlogged roots have epinastic leaves. Senescent tissues, including ripe fruit, generally produce high amounts of ethylene, so senescence that is started by ethylene is autocatalytic. The mechanism by which ethylene synthesis is regulated is not entirely clear, but it generally involves controls on the rate of ACC synthesis.

Plants have another molecule that regulates growth and development, one quite different from the hormones described previously. This molecule, phytochrome, is involved in many of the responses of plants to light.

Phytochrome is a protein attached to a light-absorbing chromophore (a tetrapyrrole, which is chemically related to chlorophyll and heme.) The protein has two forms, distinguished by their three-dimensional structures and by the colors of light they can absorb best. One form, $P_r$, absorbs red light (peak absorbance at 680 nm); the other, $P_{fr}$, absorbs far red light (peak absorbance at 730 nm). The two forms are interconvertible, and each can convert into the other after it absorbs a light photon (Fig. 1–10). Phytochrome is synthesized in the dark as the $P_r$ form. In the presence of light—at very low intensities—a fraction of the $P_r$ form is converted to $P_{fr}$. It is this $P_{fr}$ form that has the major effects on growth and development. The $P_{fr}$ form then decays or reverts to the $P_r$ form, so the system is capable of detecting subsequent light impulses.

A whole syndrome of phytochrome-mediated light effects can be identified by comparing seedlings grown in the dark with ones grown in red light. Light-grown bean seedlings have a straight stem, lacking the "hook" found in dark-grown seedlings. Light-grown bean seedlings have a shorter stem, and they have more and larger leaves which have developed chloroplasts. In the leaves are more carotenoids, protochlorophyll, and enzymes of photosynthesis. Other organs are also controlled by phytochrome: light promotes the germination of certain seeds and inhibits the germination of others. In adult plants, phytochrome is involved in the detection of day length. This influences the timing of flowering, dormancy, leaf senescence, leaf abscission, and spring bud break in many species.

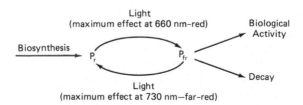

**Figure 1-10.**   Interconversion of forms of phytochrome. $P_r$ is the red light-absorbing form with peak absorbance at 680 nm; $P_{fr}$ is the far-red light-absorbing form with peak absorbance at 730 nm. In the dark, $P_r$ is synthesized. $P_{fr}$ is thought to induce physiological responses.

It is clear that phytochrome regulates many aspects of plant development, coordinating the plant to light in the environment. Phytochrome is not a hormone, in the sense that auxin and gibberellin are hormones, because it cannot move between cells and tissues. However, some of its effects may be mediated by hormones. For instance, long days and short nights, generally thought to be detected by phytochrome, stimulate the conversion of inactive forms of gibberellin to active forms in spinach. Gibberellin induces flowering in this plant. Thus, phytochrome and gibberellins may both be elements of the system by which long days stimulate flowering in spinach.

The five chemical hormones listed above are the ones that are generally discussed and studied by plant physiologists, and they are common growth regulators in all higher plants. However, they are not the only natural organic molecules that affect plant development at low concentrations. Many others are known, and some, like triacontanol and brassinolide, were discovered only in the 1970s and 1980s. In the same vein, phytochrome is not the only light receptor involved in plant development. There is, for example, a blue-light receptor, sometimes called "cryptochrome," which influences auxin transport, plastid position, and in some species, flavonoid synthesis. We are confident that information about growth regulators will accumulate, and the field will continue to appear more and more complex. But complex information is not all that is needed. What is needed is a set of principles, supported by good experimental evidence, to weld the information together. A full explanation of how one hormone influences the growth and development of one cell or one tissue could be an excellent start toward developing those principles.

## D. MOLECULAR EXPLANATIONS FOR DEVELOPMENT

One objective of developmental biologists is to explain on the subcellular level the events that cause cells and tissues to develop along specific patterns at specific times. Subcellular or "molecular" studies of development cover a wide range of topics, ranging from the sensing of a stimulus to the control of cell division and enlargement. Molecular studies may look at the production of a specific gene product which is characteristic of a certain differentiated cell. In a sense, the early studies defining plant hormones provided one level of molecular explanation for certain developmental responses. *The Molecular Biology of Plant Development* by Smith and Grierson (1982) focuses primarily on the characteristics of phytochrome, a plant molecule that senses light. Other work, equally appropriate, catalogues the various enzymes that are necessary for particular differentiated tissues to function.

Until the 1970s and 1980s, explanations of how genes operate in development were tentative and scarce. It is not that the role of genes has been unappreciated. At the beginning of the twentieth century, it was realized that an understanding of gene function would be necessary to explain development. However, technical as well as conceptual limitations held up progress. Now there are effective methods for isolating DNA, analyzing its structure, and studying its functional properties *in vitro*. Even better, DNA can be modified, so that experimental protocols can be applied to questions about the relationship between gene structure and function. Most of these new techniques were developed with bacteria and then extended to animal cells. Now almost all techniques are available to those studying plants.

Virtually all characteristics of cells and multicellular organisms depend on the expression of genes. Most characteristics come from enzymatic reactions, and enzymes are produced in response to specific instructions in the base sequences of their "structural genes" (Fig. 1–11). During development, cells in different parts of an organism acquire different characteristics. Development, therefore, represents the controlled expression of particular structural genes. Some of the more obvious gene products are listed in Table 1–2.

For a particular tissue, biologists would like to know what gene products appear and why they are necessary for the tissue's function. We also

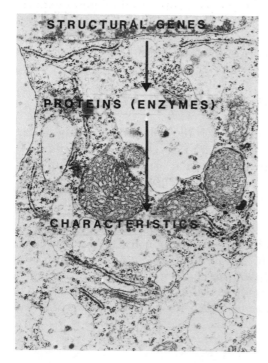

**Figure 1-11.** Flow of information in a developing plant cell. "Structural genes" are in the nucleus (part of which is seen at the top of the picture, enclosed by a double membrane). "Proteins" (enzymes) are synthesized on polysomes (dark bodies to the left and right of the upper arrow). "Characteristics" include the cell contents and macroscopic traits. Photograph supplied by Dr. John W. Oross.

TABLE 1-2    Gene Products Involved in Development.

| Stage | Event | Gene Product |
|-------|-------|--------------|
| Seed germination | Mobilization of fats | Isocitric lyase, malate synthase |
| | Mobilization of starch | Amylases, maltase |
| | Mobilization of protein | Proteases |
| Primary growth and development | Cell division | Histones, thymidine, kinase |
| | Cell elongation | Glucan synthetase |
| | Lignification | Phenylalanine-ammonia lyase, cinnamate hydroxylase coumarate-CoA ligase |
| | Chloroplast development | Ala synthetase, light harvesting chloro-plast protein, ribulose bisphosphate carboxylase, and others |
| | Cutin formation | Palmitate hydroxylase |
| Dormancy | Senescence | Chlorophyllase, proteases |
| | Abscission | Cellulase |
| Flowering | Color formation (flavonoids) | Phenylalanine-ammonia lyase, cinnamate hydroxylase, coumarate-CoA ligase, chalcone synthetase, and others |
| Seed formation | Starch accumulation | Sucrose synthase, starch synthase |
| | Protein accumulation | Prolamines, globulins, lectins, and others |

want to learn how many genes are expressed and how their expression is stimulated. With regard to the mechanism of stimulation, it would help to define not only the stimulus (for example, light, temperature, age, position in the plant), but also each step in the chain of events that leads to the induction of gene expression. And we would like to find out why other genes are not expressed.

There is a second way of relating genes to development. Certain genes modify developmental processes and produce more or less apparent changes in organs or tissues. These genes are defined by classical mating experiments. Examples are found in peas: they include the genes Mendel discovered, which govern flower color, seed shape, stem length, seed color, pod color, and position of flowers. Other genes have been defined in corn, tomatoes, and other crops (Fig. 1–12). Each alternative developmental pattern represents the allele of a particular gene. In a few cases, the gene is a known structural gene, coding for a protein, but in most cases no gene product is known and it seems likely that the gene acts by regulating other genes that do code for proteins. Such a gene would be called a "regulatory gene."

**Figure 1-12.** Mutant phenotypes in tomato. A. Dumpy (shortened internodes). B and C. Curly (foreshortened mid- and lateral leaf veins and petioles). C shows the short internode trait. D. Gold fruit (right) compared to normal red fruit (left). E. Increased ventral purple anthocyanin (right) compared to normally colored leaf (left). F. Segregation for lack of anthocyanin. Note the two light and three dark hypocotyls. The dark hypocotyls are red. Plants courtesy of Dr. C. Rick and J. Zischke.

This observation leads to the formulation of more questions. We would like to know how each given gene that affects development works. How, in particular, do regulatory genes affect the expression of structural genes in an organ? For a given tissue, we want to learn how many regulatory genes are involved in the timing and shaping of its development. These are reasonable questions, but for reasons discussed subsequently, they are at present much more difficult to answer than are questions about structural genes and their products.


## E. CHOOSING EXPERIMENTAL SYSTEMS

In experimental biology, choosing a system of inquiry can be one of the most important steps, one which determines whether your research will answer your basic questions or stagger along in confusion and frustration. There are many examples of good systems. In genetics, Mendel's pea plants were diploid and had a reasonable number of independent chromosomes. Mendel might have chosen hexaploid wheat, or *Happlopappus gracilis* with n = 2 chromosomes, or dandelions, which reproduce parthenogenically and show totally maternal inheritance. In embryology, sea urchin eggs and sperm are easy to obtain and culture; the eggs are large and easy to manipulate. In molecular biology, bacterial viruses grow rapidly, have manageably few genes, and can be purified easily. Systems for pioneering research tend to be simple, allowing the researcher to concentrate on a few techniques and a few critical questions.

The application of molecular biology to plants and their developmental processes involves some special problems. The small size of specialized tissues, the way in which tissues are firmly and intimately connected in plants, and the presence of compounds that interfere with the purification of genes and proteins all present difficulties. The rapid advances that occurred in the 1980s have often depended on correct choices of experimental material. Often the choice was dictated by the simplicity of the tissue. Sometimes the choice has been determined by the ease of experimental control over a developmental process. More recently, the choice of experimental material has been based on the expanding knowledge of genetic and physiological processes in a few well-studied species.

When the question to be answered concerns the appearance of gene products, it is useful to find a tissue that makes large amounts of those products. In animals, reticulocytes, which produce large amounts of hemoglobin, have been a favorite system. In plants, the developing seeds of grasses and of legumes are good examples. These produce large quantities of proteins that store nitrogen to be used when the seed germinates. Developing leaves are also useful. These make large amounts of certain enzymes involved in photosynthesis.

Sometimes, it is not so necessary that the gene products be present in large amounts. If conditions can be manipulated in such a way that the gene product will be synthesized exclusively or preferentially over a short period of time, then large amounts of gene products are not essential. Heat shock causes plants, like other organisms, to form a few proteins to the exclusion of the many proteins they normally make. Anaerobiosis does the same thing. The addition of certain hormones to the correct tissues can direct the genetic machinery of those tissues along certain lines.

When the question involves the genetic control of a developmental process, but the gene product is not known, then it is useful to find mutants. The genes and the gene products of mutants and wild types can be compared in the hope that the differences will explain how the gene has been changed and how it works normally. In such a study, it is necessary to know as much as possible about the genetic structure of the plant, and it is helpful to have a stock of mutants in known genetic backgrounds. There are only a few species that have been extensively studied in this way: maize, peas, and tomatoes, and (to a lesser extent) wheat, rice, soybeans, and snapdragons. Many projects in developmental botany are now focusing on these plants.

Tissue and cell culture is sometimes useful as a source of material. Pieces of plant tissue excised from the stem, leaf, or root can grow if placed in a medium that contains the correct nutrients, including carbon, sources of energy, nitrogen, salts, vitamins, and hormones (see Section A).

Culture techniques, properly used, have the following advantages for developmental studies. First, callus and suspended cell aggregates are simple tissues, containing relatively few cell types. They are grown axenically, without contaminating bacteria or fungi. Second, tissue and cell cultures are easily manipulated. They require little space, so it is easy to adjust temperature and light conditions. And they grow in intimate contact with the medium, so it is easy to add hormones or other organic or inorganic substances. Third, sometimes it is possible to induce controlled differentiation of cells in cultures (Fig. 1–13).

There are certain disadvantages to using cell and tissue cultures. First is the known heterogeneity of many cultures and the unusually high mutation rates of cells in culture. It is generally difficult to separate different types of cells in culture, because they are not in predictable positions. Second, the artificial nature of cultures leads some workers to question whether developmental processes that operate in culture are relevant to the life cycle of a plant in nature.

Many studies on plant development employ germinating seeds and seedlings. Many of the same reasons listed for tissue and cell culture apply to studies with seeds and seedlings. Seeds can be sterilized with disinfectant to exclude contaminating bacteria and fungi. Seed germination can be triggered with fair precision by the addition of water (or by light or

**Figure 1-13.** Regeneration of plantlets of *Papaver somniferum* (opium poppy) in culture. Culture prepared by Dr. A. Levy.

hormone treatment). This insures that a large amount of tissue in the same stage of development is ready when needed. There are many interesting developmental events that occur in the germination of a seed and the growth of the seedling. These include rapid cell division and elongation, mobilization of storage materials, and the initial synthesis of the biochemical pathways for photosynthesis, nitrogen assimilation, and other aspects of autotrophic life. Seeds and seedlings often (but not always) have minimal amounts of the phenolic compounds that destroy enzymes in extracts. Of course, seeds and seedlings are complex and heterogeneous, so that care must be taken to isolate the tissues in which one is interested.

There are some stages of development that occur only in a mature plant. These include flowering, accumulation of storage materials (as in seed synthesis and storage-root growth), and fruit development. In these situations, it is common to use explants, sections of the plant cut off and placed on supporting and nourishing medium. This allows preliminary selection of the tissue of interest and additions of hormones and other factors. One should always remember, however, that cutting off a plant part involves making a wound, and plants respond to wounding in characteristic ways, including the production of ethylene. And ethylene, as a plant hormone, affects many types of development. This may change the pattern of development that one is trying to study.

# TWO

# Gene Expression in Eukaryotes

In order to see how traits may be controlled in a developing plant, we must understand how genetic information is expressed. How might this expression be regulated by turning on or off one or more steps in the expression process? Our information about the principles of gene expression goes back to the 1950s, but in the 1980s there were major discoveries that helped to present a more detailed description with more concrete hypotheses about how regulation works.

In this chapter, we describe first the genetic system of prokaryotes. There are several reasons for this: (a) In general, the process of gene expression is better understood in prokaryotes, and many of the principles of gene expression were first discovered with prokaryotic organisms. This makes prokaryotes good models for both teaching and learning. (b) While prokaryotes do not develop in the sense that higher plants and animals do, they do show complex regulation of their genes. Much is known about the way they do this, and this information serves to generate hypotheses about how eukaryotic genes are controlled. (c) Certain aspects of the prokaryotic system are found in mitochondria and plastids, so some information about prokaryotes is directly applicable to plant cells.

Next we describe the genetic system of eukaryotes, emphasizing the extra complexities that arise in the flow of genetic information from nucleus to cytoplasmic trait. We will also consider how this flow could be regulated and what evidence we would need to determine the mechanism of regulation that operates in any given situation.

## A. THE PROKARYOTIC MODEL

In prokaryotes, the genome (the collection of all the genes in the cell) resides primarily on one large, circular molecule of DNA, though there can be smaller, independent DNA circles, called plasmids, which also carry genes. Each gene, including structural genes that code for proteins, consists of a contiguous sequence of base pairs. Generally, adjacent genes are separated by some number of base pairs that do not code for any product but may help control expression of nearby genes. In prokaryotes, evolution has tended to keep the genome simple and as small as possible. Thus, for most genes in prokaryotes there is only one copy on the genome. Exceptions are found for the ribosomal RNA (rRNA) and transfer (tRNA) genes. In *Escherichia coli,* there are seven copies of each of the rRNA (18S, 28S, and 5S) genes and about 60 tRNA genes (that is, between two and three times the number of different tRNAs). Often, though not always, genes that code for enzymes in one biochemical pathway are adjacent to one another and are expressed together. These groups of functionally related genes are called operons.

The expression of a structural gene occurs in two main steps: transcription and translation. In the process of transcription, the base sequence of the gene is copied onto an RNA molecule. In the process of translation, the RNA base sequence is used to direct the polymerization of amino acids to form a polypeptide chain. There may also be some post-translational processes required for the gene product to become fully active (Fig. 2–1).

Transcription starts when a molecule of RNA polymerase binds to DNA. This occurs at regions called promoters. In prokaryotes, promoters vary in base sequence but generally have two regions. One region (called the Pribnow box or the -10 region) is located about 10 base pairs upstream from the start of transcription and has a base sequence approximating TATAAT. Another promoter region is located about 35 base pairs upstream from the start of transcription and has a sequence approximating TTGACA. Note that DNA is double stranded, and *in theory* either strand could be used as a template for RNA synthesis. *In fact,* each strand is used as a template for one or more genes. But for any given gene, only

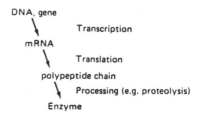

**Figure 2-1.**  Flow of genetic information in prokaryotes.

one strand is generally used. The promoter directs not only the starting position but also the choice of strands (and, as a consequence, the reading direction). RNA polymerase recognizes and binds to the TTGACA and TATAAT sequences and opens the DNA near or at the TATAAT sequence, thus exposing the bases of the template strand (Fig. 2–2).

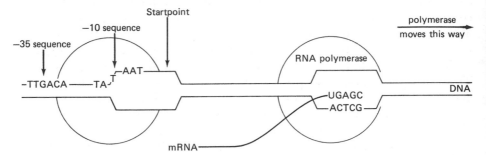

**Figure 2-2.** Association of RNA polymerase molecules with a gene promoter ("-10 sequence" and "-35 sequence").

Once the template strand is exposed, the RNA polymerase can start binding and polymerizing nucleoside triphosphates (NTP), matching each NTP to the base of the template before it is added to the growing RNA chain. As the polymerase moves, the DNA template returns to its double-stranded form and the RNA sticks out away from the DNA. The RNA remains attached to the polymerase-DNA complex by its growing end until the transcription process is completed.

Transcription continues through one or more genes and further until the RNA polymerase reads through a termination signal. Termination signals are not well understood, but they seem to share the property of having an inverted repeat in their base sequence. This may allow the RNA to form a hairpin (Fig. 2–3), and it is thought that this may be the agent that effects termination. Whatever the agent, the RNA polymerase tends to slow its activity at the termination signal, and this leads to the dissociation of the RNA-DNA-polymerase complex. For some genes, a rho factor protein (a specific termination protein) is involved in the termination. For other genes, the rho factor may not be necessary.

In prokaryotes, the process of translation generally begins while transcription is occurring, using the mRNA as a template while it is being formed. This is possible because the first part of the mRNA to be synthesized (the 5' end) is the first to bind ribosomes and be translated. The combined transcription-translation process allows electron microscopists to see structures such as those illustrated in Figure 2–4, in which DNA, RNA polymerase, mRNA, ribosomes, and growing polypeptide chains all take part.

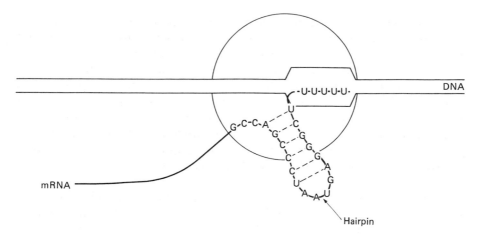

**Figure 2-3.** Postulated "hairpin" formation in RNA, thought to be related to termination of transcription.

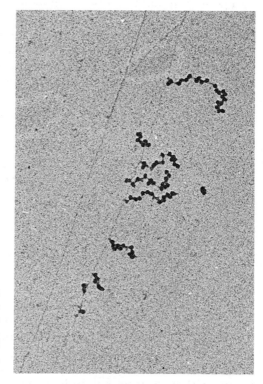

**Figure 2-4.** Combined transcription and translation of a prokaryotic gene. Reprinted with permission from O.L. Miller, Jr., B.A. Hamkalo, and C.A. Thomas, Jr., "Visualization of bacterial genes in action," *Science* 169:392–395, copyright 1970 by AAAS.

The initiation of translation is a multistep process. First, the small (30S) subunit of the ribosome binds to the mRNA. This complex is then joined by a special initiator tRNA, formylmethionine (fmet)-tRNA, together with the large (50S) subunit of the ribosome. Three initiation-factor proteins are required for the various steps of initiation, and GTP is bound and hydrolyzed during the process. The initial binding of the 30S subunit to the mRNA probably involves base pairing between the rRNA of this subunit and a base sequence on the mRNA (the Shine-Dalgarno sequence) that approximates AGGAGG. This sequence is a few bases from the trinucleotide codon, AUG, that binds to the anticodon of the initiator tRNA.

After initiation, the ribosome elongates the polypeptide chain by connecting it with peptide bonds to new amino-acyl tRNAs, one at a time. The process follows these steps (Fig. 2–5): (a) An amino-acyl tRNA with an anticodon complementary to the next codon on the mRNA binds to the ribosome-mRNA complex. (b) The fmet, or in subsequent steps the polypeptide chain, is transferred from the old tRNA to the amino group of the amino acid on the new amino-acyl tRNA. (c) The ribosome moves one codon (three bases) forward on the mRNA, bringing a new codon into position to bind to a new amino-acyl tRNA. In this process, two elongation factor proteins are needed, one in step (a) and one in step (c). In each of these two steps, a GTP is bound and hydrolyzed.

At the end of the mRNA coding sequence, the process of translation is terminated. This occurs at the termination codons, UAA, UGA, and UAG. When the ribosome reaches such a codon, one of two release factor proteins catalyzes the hydrolysis of the bond between the polypeptide chain and the last tRNA, releasing the chain. The mRNA-tRNA-ribosome complex then falls apart.

If the mRNA contains the sequences for more than one gene (that is, if it were transcribed from a polygenic operon), then separate processes of initiation, elongation, and termination of translation occur at each gene.

When transcription terminates and the mRNA is released from the DNA, the process of translation continues for a time. However, the presence of RNAase activity in the cytoplasm tends to break the mRNA, rendering it inactive as a template for translation. On the average, a bacterial mRNA has a half-life of one to two minutes and serves as a template for about 100 proteins.

There may be several steps following translation that are required before an enzyme becomes active and thus before its gene can be said to be expressed. Some of these fall under the general heading of "processing" the polypeptide chain. They include removal of some amino acids. In *E. coli* only about half the completed enzymes have formylmethionine at their N-terminal ends, even though all are initiated with fmet during the translation process. A deformylase removes the formyl group, and one or more amino peptidases may remove one or more amino acids from the N-terminal end. This is thought to occur soon after translation begins, though the signal that allows this to occur in some proteins but not others is not known. When proteins are to be secreted outside the bacterial plasma membrane, the first few amino acids are hydrophobic. This "leader sequence" or "signal sequence" inserts itself through the membrane during translation, guiding the whole protein through the membrane. The leader sequence is generally excised after the protein has passed through. The excision is another processing step.

Other steps in the activation of the protein may include the complexing of the polypeptide chain with cofactors such as heme (in cy-

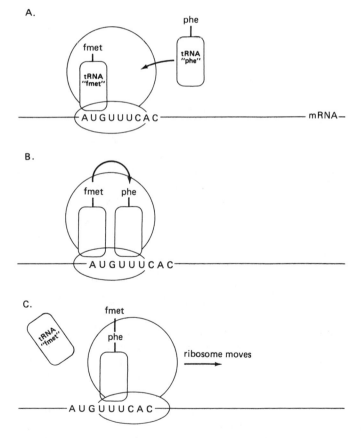

**Figure 2-5.** Elongation of the polypeptide chain during translation. A. Addition of aminoacyl-tRNA. B. Formation of a peptide bond. C. Movement of the ribosome along the mRNA.

tochromes) or pyridoxal phosphate (in decarboxylases, in isomerases, and in others). Allosteric regulation may also influence the activity of an enzyme. These steps are not usually discussed in terms of "gene expression." Since they may contribute to the channeling of biochemical activities of a cell, however, they must be considered in a general treatment of differentiation.

In theory, the rate at which a gene is expressed might depend either on the supply of mRNA (transcription-level control) or on the rate at which mRNA is used, assuming the supply is not limiting (translation-level control). Or it may depend in some cases on the rate of processing of a polypeptide chain (post-translational control). Actually, this is a simplification of the situation, since there are several possible ways to modulate the supply and use of mRNA. Nevertheless, this simple analysis allows us to say that the expression of many prokaryotic genes is controlled by the supply of mRNA and that the control of mRNA supply often involves regulating the rate of initiation of transcription. Several mechanisms for controlling transcription are now known.

The first mechanism to be discovered was repression. This is a negative control in which the binding of a repressor protein to DNA turns off the expression of a single-gene or polygenic operon. The control is specific because the repressor binds to a region of the DNA, the operator, that is close to the promoter for the operon. The repressor sterically prevents RNA polymerase from attaching to that promoter. Control can be reversed by changing the structure of the repressor. This often involves the binding of a molecule in the cell ("inducer") whose concentration indicates the need for expression of the operon concerned. The quintessential example of a repressor is the lac repressor, which binds the operator for the genes that control the metabolism of lactose (Fig. 2–6). In the presence of the inducer (a metabolite of lactose), the repressor stops binding to the operator, and the cell begins transcription and translation of the genes for lactose permease and beta-galactosidase. The gene products then allow the cell to use external lactose as food. Another repressor is the tryptophan (trp) repressor. In the presence of trp (a "corepressor"), this repressor binds to trp and to the operator for the genes that control the synthesis of trp. When the concentration of trp in the cell falls, trp leaves the repressor binding site, the repressor leaves the operator, and transcription and translation of these genes begin. In general, each repressor fits one operator in the genome, and the binding of each repressor is influenced by one specific compound (lactose metabolite, trp, and others). This means that the presence or absence of the compound in the cell controls the expression of the operon.

The second mechanism involves the CAP protein (cyclic AMP-binding protein). This is a positive control mechanism which turns on several genes in the cell (Fig. 2–7). The CAP protein is activated by the com-

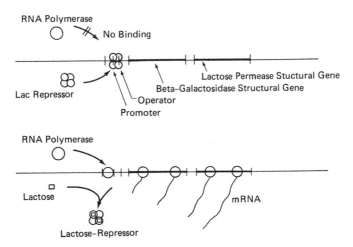

**Figure 2-6.** Control of the lactose (*lac*) operon by repressor. A. Repression in the absence of lactose. B. Induction by the addition of lactose and consequent change in DNA-binding ability of the *lac* repressor.

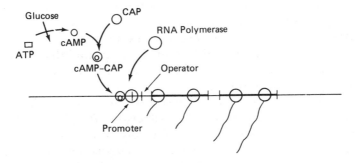

**Figure 2-7.** Control of the lactose operon by cyclic adenosine monophosphate (cAMP) and cAMP-binding protein (CAP). Together cAMP and CAP stimulate initiation of transcription. Glucose inhibits the formation of cAMP.

pound, cyclic AMP, which in *E. coli* is formed when the cells run short of glucose. When activated, this protein binds upstream from the promoters of the lac operon, the gal operon (genes for the metabolism of galactose), the ara operon (genes for the metabolism of arabinose), and other operons. It stimulates the binding of RNA polymerase to the promoters of these operons. It is not known exactly how the protein acts, but the promoters concerned seem not to have ideal base sequences around the -35 region. Perhaps the CAP protein substitutes for these sequences.

A third type of control is exerted by a subunit of RNA polymerase itself. This subunit, the sigma factor, acts as an initiation factor (Fig. 2–8). Without sigma factor, RNA polymerase "core enzyme" binds to DNA, but it binds at random places and does not begin transcription (does not ex-

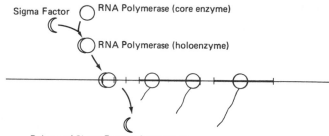

**Figure 2-8.** Control of initiation of transcription by sigma factor. Sigma factor promotes the specific binding of RNA polymerase to promoters.

pose the template strand) when it does bind. When sigma factor forms a complex with the core enzyme, the enzyme binds less well to random places on the DNA but it binds better to promoters. It also starts transcription when it binds. In practice, there is a cycle: the sigma factor binds to free core enzyme; the complex finds and binds to a promoter; the transcription starts; and almost immediately the sigma factor is released. The special function of sigma factor in initiation raises the possibility that it might recognize some promoters better than others and thus control the relative transcription rates of different genes. Even more interesting is the possibility that there might be different sigma factors that recognize different sets of promoters. This could lead to an endogenous program of differential gene expression.

An example of sigma factor control is found in sporulation—a type of bacterial differentiation—of the bacterium *Bacillus subtilis*. During growth in culture, *B. subtilis* has an RNA polymerase with a normal four-subunit core enzyme and a sigma factor (sigma-55) with a molecular weight of 55 kD ("kD" stands for kilodalton, or 1,000 atomic mass units). When nutrients become scarce, this bacterium goes through a complex process that leads to its DNA being encased in a small hard-walled spore, which is inside and at one end of the original cell. The new gene products that are required for this process are transcribed by an RNA polymerase that has the normal core enzyme joined together with a new and smaller sigma factor (sigma-37). The details of the appearance of sigma-37 are not known, nor are its effects, but we presume that the new sigma factor enables the RNA polymerase to read a new set of promoters. A second stage in the sporulation process comes when one of the early sporulation gene products (sigma-29) allows RNA polymerase to read a third set of promoters. Thus, changes in the specificity of RNA polymerase, directed by the presence of particular sigma factors, provide a positive control mechanism that turns on a large set of genes in the coordinated fashion needed for a developmental process.

In the previous four paragraphs, we described how the supply of mRNA may depend on the rate at which transcription is initiated. The supply of mRNA may also depend on the rate at which transcription aborts, that is, fails to continue through the structural genes. A process like this, called attenuation, occurs in operons that control the synthesis of various amino acids, like trp. In the trp operon, the transcription of the mRNA may terminate prematurely (attenuate) around base 140. This occurs in the presence of trp, and it is thought to occur because a ribosome reading a small gene at the beginning of the mRNA allows the mRNA to form a termination hairpin signal. In the absence of trp, the ribosome cannot move into position, and the hairpin cannot form.

Most of the processses of gene expression in prokaryotes occur in eukaryotes. It has been hypothesized that some of the mechanisms by which gene expression is controlled in prokaryotes may also occur in eukaryotes. As we will see in the next section, there are many other possibilities for control in eukaryotes.

## B. THE EUKARYOTIC SYSTEM

The genome of eukaryotes differs from that of prokaryotes in many ways. In general, it is larger. While bacterial genomes range from $10^6$ to $10^7$ base pairs, plant genomes have from $10^8$ to $10^{11}$ base pairs. It is also more complex and more scattered. For instance, it is divided by membranes into several spatially separate systems: nucleus/cytoplasm, mitochondria, and in plants, plastids. In each system, there are unique genes that code for mRNAs and unique machinery for synthesizing proteins from these mRNAs.

In the nuclear system, which contains by far the largest amount of DNA and the largest number of genes, the genes are divided among several independent chromosomes. There may be one to over one hundred chromosomes, and even on one chromosome, the genes are distributed in a less compact manner. They are often separated from one another by long stretches of DNA that have no obvious function. Individual genes are also commonly divided into segments (exons) that are separated by noncoding DNA (intervening sequences, or introns). There are no operons—no groups of functionally related genes that are transcribed together.

In the nucleus there are often multiple copies of base sequences. Certain genes that code for well-known proteins, like globins and immunoglobulins, are members of gene families in which there are many copies with very similar base sequences. Many of these genes, though not necessarily all, will be transcriptionally active. Other base sequences—some long and some short—are present in very many copies, yet they may not be transcribed. The function of these sequences is not known. While they may help regulate the expression of other genes, they may be simply "spacer" or "junk" DNA with no essential purpose.

The physical structure of the nuclear genome also differs from that of prokaryotes. A different physical structure is necessary because a very large amount of DNA must be packed in a small space. For example, in rye plant cells 16 pg of DNA, which if extended would run five meters in length, must fit in a nucleus 5 micrometers in diameter.

Nuclear genes are found in chromatin, which is DNA complexed with histones and some non-histone proteins. The DNA winds around a globular aggregate made of two molecules each of H2a (histone subunit 2a), H2b, H3, and H4 to form a nucleosome. The nucleosomes are spaced about 200 nucleotides apart. Under some conditions, an electron micrograph of chromatin (Fig. 2–9) shows the nucleosomes in an extended conformation. In this conformation, the nucleosomes look like beads on a string.

While chromatin may be extended in some regions of the chromosomes at some times during the cell cycle, the chromatin can be condensed and have higher orders of structure. This helps it pack into a smaller space. Condensed chromatin is thought to have more complex helical coiling, which is probably controlled by associations between the nucleosomes. Such condensed chromatin can be seen in the condensed chromosomes of mitosis and meiosis, in the polytene chromosomes of fruit flies, and in some dark-staining heterochromatic regions of interphase nuclei (such as condensed X chromosomes or "Barr bodies" in mammalian females). It is not necessarily true, however, that each of these forms has the same type of condensed structure.

The eukaryotic system for gene expression is in many ways like the prokaryotic system, but there are extra complications. These complications result from (among other things) the packing of DNA in chromatin; the splitting of gene segments by introns; and the localization of the DNA in the nucleus, away from the site of protein synthesis. These considerations mean that there will be extra steps that must occur in their proper sequence before a genetic trait can be observed and measured. Each new step is important because it represents a possible control point at which the expression of a gene may be turned on or off. Figure 2–10 shows some of the steps needed for the expression of eukaryotic genes.

It seems likely that DNA in the highly structured forms of chromatin, and possibly also in the extended string-of-beads form, is not capable of serving as a template for transcription. Packing of DNA and histones in a dense mass would prevent the approach of RNA polymerase, and the coils around the histone aggregates might interfere with the separation of the DNA strands. This suggests that a first step in gene expression is to prepare the template sequences for transcription, a step that we

**Figure 2-9.** Chromatin structures. A. Diagram of a nucleosome. DNA makes 1.8 turns
around a core of histones. A te-
tramer, containing two subunits
each of H3 and H4, binds to the
central turn of the DNA, while
dimers, each with one subunit of
H2A and H2B, sit at the front
and at the back. B. 10-nm fiber,
showing nucleosomes. C. 30-nm
coiled structure.    Photographs
courtesy of Dr. Barbara A.
Hamkalo.

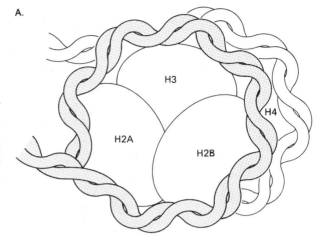

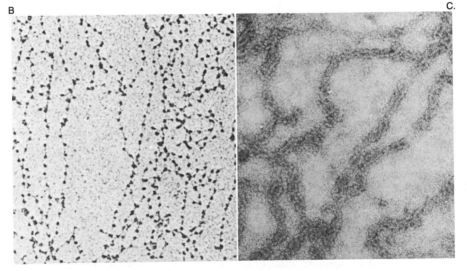

will call "activation." A visible example of activation may be found in the
puffs of insect polytene chromosomes. In contrast to the major part of the
chromosomes, these puffs are transcriptionally active. They have an appearance and staining behavior that suggests a different physical structure from the inactive regions. More evidence for activation comes from
studies of nuclease sensitivity of chromatin. There are small regions that
are preferentially digested by DNAase I, an enzyme that hydrolyzes de-
oxyribose-phosphate bonds in double-stranded DNA. It seems likely that
these regions are more open and thus more exposed to the enzyme. Some
of these regions have been associated with active gene expression. In
chicken red blood cells, globin genes (which are actively expressed) are six

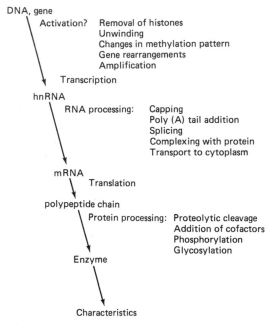

**Figure 2-10.** Expression of genetic information in the eukaryotic nucleus. In-formation contained in the base sequences of DNA must be made available for transcription ("activation"), then transcribed onto hnRNA (or nRNA, "nuclear RNA"). The nRNA must be processed before it becomes mRNA. Processing involves both chemical modifications and transport to the cytoplasm. The mRNA is translated to form a polypeptide chain. This polypeptide chain may need processing through chemical modification before it acquires its functional activity.

times more sensitive to DNAase I than are ovalbumin genes (which are not expressed).

We can hypothesize that genes must be activated to be transcribed. And we can show certain changes in their physical structure. But we do not really know whether activation precedes or follows the beginning of transcription. Also, we have no certain knowledge of the steps that are required to effect activation. The first obvious possibility is a change in the aggregation of histones—this may be controlled by acetylation or phos-phorylation of the histone subunits. The second possibility is that changes in the types of histones that are present may provide tissue specificity. For example, in sea urchins, different subtypes of histone subunits appear at different developmental stages. A third possibility involves changes in nonhistone proteins that bind to DNA and alter its ability to associate with histones.

Activation of nuclear genes is influenced by certain base sequences which are called "enhancers." An enhancer was first discovered in the genome of the virus SV40. This 72-base pair sequence had the remarkable property of increasing RNA synthesis if it were present anywhere within a thousand or so base pairs of the gene. Enhancers have now been located near cellular genes (immunoglobulin genes). While it is not clear what they do, it is possible that they help put the adjacent genes into a physical state appropriate for transcription. Enhancers have been associated with two other physical characteristics of DNA: the Z helical form and methylation.

Z DNA, a left-handed helical form of DNA (Fig. 2–11), was discovered by x-ray crystallography of the dry crystals of synthetic polydeoxyribonucleotides, but it has also been detected in short stretches of DNA in the nucleus. It is present in certain enhancer regions. It is most stable in regions of alternating purine-pyrimidine base sequence and in the presence of polyamine molecules. It can be stabilized by methylation of certain bases. This is interesting, because for many years methylation has been implicated in differentiation.

In animal and plant DNA, C bases may be chemically modified by the addition of a methyl group. In animals 2% to 7% of the C bases are methylated, and all these modified C bases are adjacent to G in CG doublets (Figure 2–12). In plants a much larger percentage (over 25%) of the

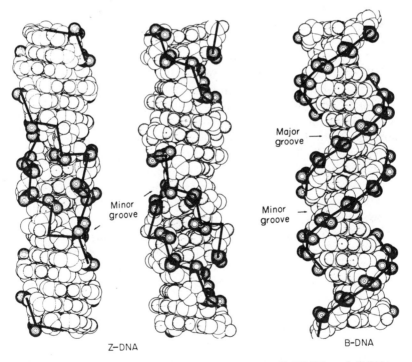

**Figure 2-11.** Comparison of the helical structures of Z-DNA and B-DNA. Reprinted by permission from Wang and others, *Nature* 282:680 Copyright © 1979 by Macmillan Journals Limited.

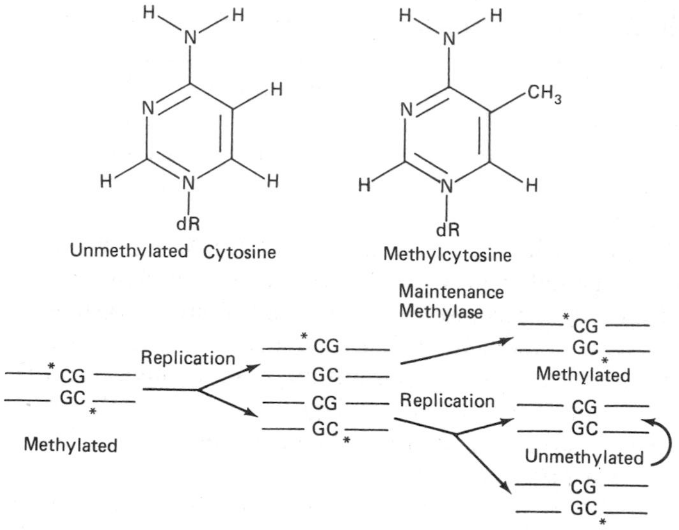

**Figure 2-12.** Methylation of cytosine bases in DNA. A. Unmethylated cytosine (dR, deoxyribose). B. Methylated cytosine. C. Disposition of methylated cytosine after replication of DNA. Alternative events include maintenance of methylated CG pairs (upper fork) and generation of an unmethylated CG pair by a second replication step (lower fork).

C in DNA is methylated: modified C bases are found both in CG doublets and in CNG triplets (N is another, unspecified base). In both cases there is a symmetry between the two strands of DNA. A CG doublet will be opposite another CG doublet, and a CNG triplet will be opposite another CNG triplet, although the Ns will differ in the two strands. An enzyme called maintenance methylase acts to preserve an existing methylation pattern by methylating any C in a newly synthesized CG doublet that is opposite an already methylated CG doublet. If this enzyme is interfered with, however, after two rounds of replication an unmethylated CG/GC will exist. And it will be stably unmethylated in the presence of the maintenance methylase. Initiation methylases can produce a new set of methylated CG doublets. This situation has possibilities for controlling development, because it clearly produces stable, yet reversible, changes in the genome. The evidence linking methylation with differentiation is indirect and contradictory. As mentioned above, methylation has been connected with Z DNA and enhancers which promote gene expression. Yet methylated DNA has a higher melting temperature than unmethylated DNA. This suggests that methylation would hinder transcription. In addition, methylated DNA is found most often associated with inactive DNA regions (satellite DNA, "Barr bodies," centromeres). The genes for beta-globin, one of the protein subunits of the hemoglobin molecule, are methylated more in certain tissues, such as brain and sperm, than in others, such as liver and cultured cells. But none of these tissues synthesizes the globin molecule. One can only hypothesize that methylation has some role in development—more data will be needed to define that role.

There are other changes in the genes that may precede transcription in differentiating systems. One involves rearrangements. These are seen most clearly in cells that produce immunoglobulins (antibodies). Immunoglobulins are proteins that have binding sites that stick to other molecules. Each immunoglobulin molecule is made of four polypeptide chains: two "light" chains and two "heavy" chains. The type of binding site is determined by the amino acid sequence of the light and heavy chains. All vertebrate animals have hundreds of thousands of cells, each of which makes a different light chain and heavy chain and thus a different immunoglobulin with a different binding site. This diversity is generated, in part, by rearrangement of gene segments. In the development of these cells, the gene for the light chain is spliced together from one of a family of V-gene segments and one of a family of J-gene segments (Fig. 2–13). The gene for the heavy chain subunit is spliced together from one member of each of three gene families. Because different members of the segment families can be used, and because there is some randomness in the points of joining, each stem cell gets a different gene from the splicing and thus a different immunoglobulin. Although this developmental process is unusual in the degree of randomness, it is not unique in involving gene rearrangements. Yeast cells change their mating types by movement of genes.

Another process related to activation is amplification. In this process, the genes are replicated many-fold, so that there are multiple templates available for transcription. As is found in gene rearrangement, this is not a common process, but several examples are known. One example involves the genes for ribosomal RNA in the macronucleus of *Tetrahymena,* which are multiple copies of the genes in the micronucleus. Another ex-

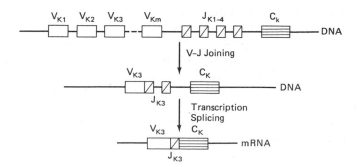

**Figure 2-13.**   Rearrangement (and subsequent expression) of immunoglobulin light chain genes. Combining different $V_k$ and $J_k$ genes gives different coding sequences.

ample is the adaptation of animal cells to the antibiotic methotrexate: Amplification of the gene for dihydrofolate reductase (the enzyme inhibited by methotrexate) provides more template and thus more mRNA and enzyme. With additional enzyme the cells develop some resistance to a given amount of the antibiotic. In yet another example, amplification of an esterase gene, which leads to high levels of a detoxifying enzyme, makes mosquitos resistant to organophosphorus insecticides.

Transcription of activated genes in eukaryotics occurs through the action of RNA polymerase, as it does in prokaryotes. In eukaryotes, there are three nuclear RNA polymerases, each a large, multi-subunit enzyme. RNA polymerase I synthesizes only ribosomal RNA; RNA polymerase III synthesizes tRNAs and some other small RNAs. RNA polymerase II is responsible for the synthesis of RNA from structural genes. Promoter regions for RNA polymerase II have been estimated from a comparison of the base sequences upstream from many different eukaryotic genes. There are two "consensus" regions that are good candidates for promoters. They have base sequences similar to those of prokaryotic promoters. Direct evidence that these regions are actually promoters has come from mutational studies, in which small changes inserted in these sequences decrease transcription *in vitro*. The first region, known as the TATA (or Goldberg-Hogness) box, has approximately the sequence TATAAAA and is located at about position -25, that is, 25 base pairs upstream from the point at which transcription starts. The second region, known as the CAAT box, is more variable and is missing from some genes. When present, it has approximately the sequence GGCCAATCT and is located somewhere near positions -70 to -80.

Once transcription has begun, it is likely that the RNA polymerase works like that in bacteria, though the details are less well known. In addition, the details of termination are less well defined. Termination signals of eukaryotic genes read by RNA polymerases I and III seem to be poly(U) sequences embedded in GC-rich regions. This is the same as for prokaryotic termination signals. The termination signals for eukaryotic genes read by RNA polymerase II have not yet been determined.

The RNA that is the product of transcription is not immediately useful as mRNA. It is more properly called nRNA or hnRNA (nuclear RNA or heterogeneous nuclear RNA). This reflects the fact that this RNA must pass through another set of steps, called "RNA processing," before it can be translated (Fig. 2–14). Most of the hnRNA is unstable and quickly breaks down in the nucleus. Only a small fraction is selected for processing. Despite the importance of the selection in the eventual expression of the genetic information contained in the RNA base sequence, it is not known how that fraction is selected. We do know that the selection is not random.

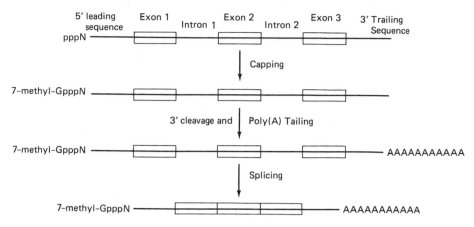

**Figure 2-14.** Processing of hnRNA (nRNA) to make mRNA. Thick blocks represent exons (coding sequences). Thin lines represent introns (intervening sequences) and 5' leading and 3' trailing sequences on the RNA. See text for description of processing steps.

The first step in RNA processing is capping, the addition of a GTP to the 5' end of the RNA by the enzyme guanylyl transferase. This GTP is connected so that its 5' end is attached to the triphosphate at the 5' end of the RNA, thus: G(5')ppp(5')NpNp. (N stands for any ribonucleoside.) As part of capping, methyl groups are added to the 7-carbon of the terminal G and sometimes to adjacent bases. A second processing step involves cleavage of the nRNA to form a new 3' end. The 3' end is then elongated with a chain of adenosines. A sequence AAUAAA is located six to twenty-six bases from the cleavage sites of almost all animal cell nRNAs, and this sequence is required for efficient cleavage. Polyadenylation, catalyzed by the enzyme poly(A) polymerase, generally follows cleavage but does not specifically require the AAUAAA sequence. Seventy percent of the RNAs destined to become messengers get a poly(A) tail, which is called poly(A)+ RNA. A third processing step is related to the genetic structure of eukaryotic DNA. When coding sequences (exons) are separated by introns, the whole assembly (exons plus introns) is transcribed onto one nRNA. Then the introns are removed and the coding sequences spliced together. The mechanism of this reaction, and the way in which the intron-exon junctions are so carefully delimited, was a mystery when the process was first discovered. Then it was discovered that ribosomal RNA from *Tetrahymena* catalyzes the removal of its own intron without the aid of an enzyme. It thus seems possible that a similar process occurs with nRNA in the process of becoming mRNA. Other steps in the processing of nRNA include a complexing with several proteins and small RNA molecules found in the nucleus to form a body called a "spliceosome."

The proteins and small RNAs probably hold the ends of the RNA together during the splicing process, and they may contribute to the accuracy and specificity of the process. Eventually, the RNA is transported out of the nucleus and into the cytoplasm, where it can be translated.

There is considerable evidence that cells regulate the RNA processing steps so that only subsets of nRNAs become mRNAs. The subset differs in differentiated cells of various types. Liver and kidney tissues in mice have the same base sequences in their nRNAs but different subsets of base sequences in their mRNAs. The same thing can be said of tobacco plants. In tobacco plants stem and leaf tissues both have nRNA base sequences, which total about $10^8$ bases; about three-quarters of these base sequences are present in both tissues. The base sequences in the mRNAs of stems and leaves total only about $3 \times 10^7$ bases, and very few of these sequences are present in both tissues. Therefore, most of the nRNA that is translated from active DNA templates is broken down, and only a selected fraction is processed into mRNA and transported to the cytoplasm.

Of course, if there are control signals that influence translation, the mRNA may not be translated even if it reaches the cytoplasm. One possible alternative fate for messenger, albeit temporary, is to be complexed with protein to make a "ribonucleoprotein" (RNP) and stored in an inactive state. This "masked messenger RNA" was originally discovered in sea urchin eggs. In sea urchin eggs pretranscribed and preprocessed mRNA is stored ready to be translated following fertilization. A similar, protein-complexed mRNA has been found in wheat-seed embryos. Conceivably, it serves a similar function in the early stages of germination.

Another alternative fate of mRNA is decay. An mRNA will eventually be broken down by nucleases in the cytoplasm. In general, mRNA in eukaryotes lasts much longer than in prokaryotes (hours instead of minutes), but the rate of breakdown varies widely, and sometimes specifically, for certain mRNAs. For instance, the stability for the mRNA for nitrate reductase in cultured plant cells varies by as much as two-fold depending on the presence or absence of a source of reduced nitrogen for the cells' nutrition.

The process of translation in eukaryotes is very similar to that in prokaryotes. The initiation of the process involves the ordered formation of complexes among an initiation factor protein (eIF2), GTP, the initiating methionyl-tRNA (met-tRNA), and the small (40S) ribosomal subunit. This complex, along with other initiation factors, binds to the 5' end of an mRNA and then moves along the mRNA until it reaches the first AUG codon. (This is in contrast to the prokaryotic system, in which the ribosome-tRNA complex binds directly to the AUG codon.) The large (60S) ribosomal subunit joins the complex along with the hydrolysis of the GTP and the release of the initiation factors.

The elongation and termination steps of translation occur in the same order as in prokaryotes, and they seem to involve similar protein factors. However, this may be a very superficial view—there may be significant differences in the mechanisms through which the two systems operate.

We mentioned in the previous section that there may be post-translational protein-processing steps that are necessary before the gene product is completed. In eukaryotes, these steps are more varied and more important. One reason for this is the existence of more membrane-bounded compartments into which proteins must be directed. For instance, the proteins of lysosomes are sequestered in small, closed vesicles. These proteins are synthesized by ribosomes that attach to the endoplasmic reticulum. The attachment involves a hydrophobic "leader" or "signal" sequence of amino acids at the N-terminal end of the protein (Fig. 2–15). This leader sequence binds to a signal recognition particle (SRP) made up of six proteins and a special small (7S) RNA molecule. The SRP binds, in turn, to a receptor protein on the endoplasmic reticulum (ER). The leader sequence and the rest of the polypeptide chain is directed through the ER membrane as translation proceeds. Once the whole protein is completed, the leader sequence is cleaved off and degraded. The remaining polypeptide chain lies within the lumen of the ER. A budding of the ER forms the lysosomal vesicles with the proteins inside. Similar processes explain the storage of proteins in other organelles, such as the vacuole of plant cells and the insertion of proteins into the structure of membranes. Certain organelles formed by the ER may be modified by the Golgi apparatus. This may involve further pro-

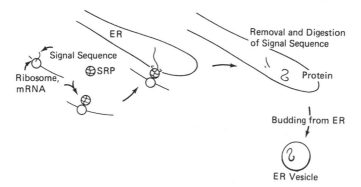

**Figure 2-15.** Translation and processing of a protein on the endoplasmic reticulum (ER). SRP is a "signal recognition particle." First it binds to the signal sequence on the N-terminal end of the protein as it is translated, then it binds to a recognition site on the ER. Further translation forces the protein into the lumen of the ER.

cessing of the proteins with, for instance, the addition of carbohydrate groups. This is thought to occur when vesicles synthesized by the ER are destined to join the plasma membrane.

Many proteins remain inactive until their polypeptide chains are cleaved by proteases (Fig. 2–16). This allows their enzymatic activity to be controlled in regard to time and space. It also allows their enzymatic activity to be controlled by changes in environment. Examples of such proteins include insulin, which has two chains cleaved from a large single chain called proinsulin. (Proinsulin, in turn, comes from preproinsulin through the removal of the latter's leader sequence.) Other examples include trypsin and chymotrypsin.

An even more spectacular example of protein processing in eukaryotes involves the synthesis of multiple gene products from one gene. The original gene product is a single polypeptide chain. The functional molecules are fragments of this chain, produced by proteolytic cleavage. Sometimes different types of functional proteins are produced by cleavage at different points in the chain, and this can depend on the tissue where cleavage occurs. The best studied example is the complex of polypeptide hormones produced by the pituitary gland (Fig. 2–17). An original gene product, called pro-opiomelanocortin is inactive, but it may

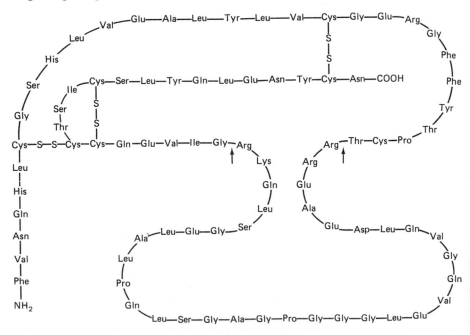

**Figure 2-16.**  Activation of human insulin. Removal of the central section (lower right) by a proteolytic enzyme (arrows) leaves two chains held together by disulfide bonds. This is the active hormone.

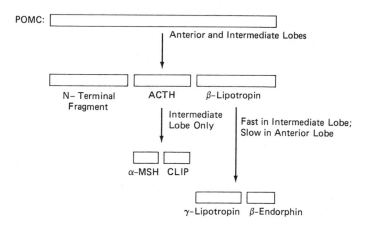

**Figure 2-17.** Processing of pro-opiomelanocortin (POMC) by proteolytic enzymes in the pituitary gland. Each box represents a polypeptide chain. Each arrow represents one or more proteolytic cleavage events occurring in the anterior or intermediate lobe of the gland. ACTH, adrenocorticotropic hormone; MSH, melanocyte stimulating hormone; CLIP, corticotropin-like intermediate lobe peptide.

be cleaved in the anterior and intermediate lobes of the pituitary to form adrenocorticotropic hormone (ACTH) and beta-lipotropin. The anterior lobe releases ACTH. In the intermediate lobe, the ACTH is cleaved to form alpha-melanocyte stimulating hormone (MSH) and corticotropin-like intermediate lobe peptide (CLIP). Also in the intermediate lobe beta-lipotropin is cleaved to form beta-endorphin and gamma-lipotropin. Clearly, in this case, development is channeled through switches at the level of protein processing.

Other types of processing involve the addition of moieties to the amino acid residues. In addition to the carbohydrates added in the Golgi apparatus, moieties such as lipids, complex cofactors, and phosphate groups may be added to proteins. Phosphorylation, in which a phosphate is transferred from ATP to a serine, threonine, or tyrosine residue by a protein kinase, is especially important, since it forms one or more steps in "cascades" of events controlling important biochemical pathways. Phosphorylation plays a part in such biochemical processes as glycogen metabolism, glycolysis, gluconeogenesis, and protein synthesis. In glycogen metabolism for instance, the activity of the enzyme glycogen phosphorylase depends on whether a key serine residue has been phosphorylated by phosphorylase kinase or whether it has been dephosphorylated by a protein phosphatase. The activity of phosphorylase kinase is itself dependent on phosphorylation and thus on the relative activities of a protein kinase and protein phosphatases.

In conclusion, it can be seen that in eukaryotes there are many steps, each of which must function properly for the successful expression of a gene. There are, therefore, many *potential* points at which gene expression may be controlled. There is also evidence that in eukaryotes there are many points at which gene expression *is* controlled. Many examples of gene activation can be found, including the special cases involving immunoglobulin gene rearrangements and amplification of rDNA. And there is evidence for differential control of transcription (though this may be indistinguishable from gene activation). Indirect evidence exists for control at the level of RNA processing, involving tissue-specific differences in mRNA that are not seen in hnRNA. Control at the level of translation occurs in reticulocytes that are deficient in heme: Phosphorylation of an elongation factor stops translation of the globin mRNA. Post-translational processing, involving proteolysis and other processes, controls the activity of numerous enzymes and the final steps in the synthesis of some vertebrate peptide hormones.

The foregoing list of examples should not obscure the fact that we know very little about the control of most genes. We certainly have no general principles on which to make predictions. For many situations in plant development, we are at the first stages, trying to decide what genes to study and at what level the control of gene expression occurs.

## C. STRATEGIES FOR STUDYING CONTROL OF GENE EXPRESSION

There are some general methods for determining the level of control of a gene's expression in a developing tissue. It is valuable to discuss these methods because some can be easily misinterpreted.

One technique uses chemical (sometimes physical) inhibitors that disturb more or less specifically a single step in the expression of a gene. This technique has been used for many years, because it is fast and easy. There are many potential inhibitors that can be used: Actinomycin D, for instance, binds to DNA and inhibits RNA synthesis. Alpha-amanitin, a product of the *Amanita* mushroom, binds to and inhibits transcription by the RNA polymerase II of nuclei. Cordycepin (3′ deoxyadenosine) inhibits the synthesis of poly(A) tails on hnRNA. Cycloheximide inhibits translation on eukaryotic cytoplasmic ribosomes, whereas chloramphenicol inhibits translation in prokaryotes (and in mitochondria and chloroplasts). These are some of the older, more familiar compounds. Many others have been discovered that often have advantages over these.

One commonly sees experiments in which an inhibitor has been added to a developing system. One may ask whether rifamycin, a transcription inhibitor, stops the induction of beta-galactosidase in *E. coli*. Does chloramphenicol stop the same function? Does alpha-amanitin inhibit the synthesis of yolk protein in frog ovaries? Does cycloheximide duplicate this same process? If the answer is *yes*, one may conclude that transcription (or translation, depending on the inhibitor) is required for the induction process. If the answer is *no*, then transcription is not required.

Inhibitors have several characteristics of which one must be aware. First, inhibitors have limited specificity. Many inhibitors are not as specific as one would like. This is especially true in experiments on intact cells, where so many processes contribute to gene expression. Cycloheximide often inhibits the appearance of an active enzyme. Generally this is attributed to its effects on ribosomal function. But a small effect of the chemical on energy metabolism may inhibit activation by phosphorylation, whereas an investigator may assume that the lack of activity reflects a lack of translation. Consequently, a positive result may be misinterpreted.

A second characteristic of inhibitors involves uptake. Many inhibitors are large, hydrophilic molecules not easily taken up into cells or into complex tissues and organs. This means that a lack of effect may occur because the chemical did not act. A positive control showing the effectiveness of the compound is needed. Otherwise an investigator may assume a negative result when in actuality the chemical did not act.

One should look carefully at the logic of these experiments. An indication that transcription (or translation) is necessary for a gene's expression does not imply that this step is blocked when the gene is not expressed. Some other process may also be necessary, and this additional process may be the step that is blocked by the control mechanism. It is more informative in an experiment if an inhibitor does not inhibit the expression of a gene that occurs as a tissue differentiates (assuming the inhibitor has entered the cells and stopped the desired process). This suggests that the process inhibited by the compound has already occurred, and that an intermediate in the pathway of gene expression has accumulated. It also suggests that the differentiation and gene expression involves the release of a block in the conversion of the intermediate to the final gene product. Unfortunately, it is unusual to find such results in experiments.

A second and more elegant technique for investigating gene control is to look for the presence of intermediates of gene expression when development is turned off. Intermediates include activated genes, hnRNA with the correct base sequence, mRNA, and polypeptide chains in the unprocessed and processed forms. One assumes that if the flow of genetic information through these intermediates is turned off at a specific point, then the intermediates below that point will disappear, whereas the intermediates above that point will be detectable. There are still problems in this strategy, however. For instance, the rapid breakdown of a particular base sequence in hnRNA represents a turning off of its processing to mRNA, but it will look (in the experiment) like a loss of synthesis. The beauty of this experimental strategy is that new, very sensitive methods for detecting individual sequences of DNA, RNA, and protein have become available.

# THREE

# Detection of Induced Gene Products

The controlled expression (or the lack of expression) of various genes represents one key aspect of development. In order to study such expression and its control, we must find systems in which the products of expression can be assayed. Because gene expression occurs in cascades in which one product leads to the formation of another, we need systems in which we could detect all intermediates, as well as the final product, of the cascade. As described in the previous chapter, this should allow us to infer the location, and maybe the mechanism, of control. The systems to be described below are cases in which the expression of one or more genes can be controlled and at least some of the intermediates can be detected. Research on the different systems has reached different levels of sophistication, but each system adds something to our knowledge of gene control in plants.

## A. HEAT SHOCK

When the cells of an organism are raised to a temperature above what is normal for them, they produce a new set of proteins, called "heat-shock" proteins. This effect seems to be true for a wide variety of organisms—it has been demonstrated with embryonic fibroblasts from chickens and CHO cells from mice. It has also been seen in *Escherichia coli*, *Tetrahymena*, *Dictyostelium*, and yeast. In the plant kingdom it has been shown in maize and soybean. The significance of the induction of the new proteins is still not clear. It is believed to be a generalized stress response, since in some organisms the same proteins may appear following exposure to anoxia, various poisons, viral infections, and other agents. Though it is not a common feature of development in a plant, we think it is interesting as one example of how environmental stimuli may change the pattern of gene expression.

The heat-shock effect was discovered by Ritossa, who was investigating chromosome puffs in *Drosophila*. These diffuse regions of polytene chromosomes are generally recognized as regions of active gene expression for one can demonstrate the presence of filamentous DNA and much RNA in the puffs. Ritossa showed that two particular regions of chromosome 2 formed puffs after forty-five minutes at 32° C. (In the laboratory, *Drosophila* normally live at 15°C to 25°C.) He showed that the same puffs appeared after one hour of anaerobic treatment and after treatment with respiratory inhibitors. He also demonstrated an incorporation of radioactive uridine into RNA at the sites (Fig. 3–1).

Later, to confirm that the puffs were actually coding for new proteins, several experiments were conducted by different laboratories. (A) Following heat treatment, radioactive amino acids given to *Drosophila* cells appeared in a small number of proteins (the heat-shock proteins). (B) It was noted that following heat treatment, the proportion of ribo-

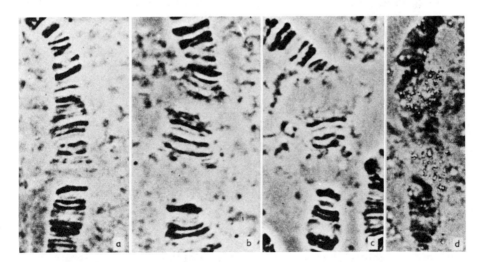

**Figure 3-1.** A section of a *Drosophila* salivary gland chromosome, showing puff formation after stress treatments. A. Normal at 25°C, third instar larva. B. 45 minutes at 32°C. C. 45 minutes in 0.01 M sodium salicylate. D. 45 minutes at 32°C, followed by 10 minutes in $^3$H-uridine. The location of the silver grains (formed autoradiographically by incorporated $^3$H-uridine) indicates that RNA synthesis is occurring specifically in the puffs. Reprinted from F.M. Ritossa, *Experimental Cell Res.* 35:601, 1964.

somes found in polysomes declined (at ten minutes), then rose (at thirty minutes). This was taken as a sign that protein synthesis had slowed, then resumed. (C) Messenger RNA was isolated from the new polysomes, and it was shown that this messenger could be translated *in vitro* to make the same proteins seen when heated cells were labeled *in vivo*. (D) Finally, the messenger RNA was shown to hybridize to the DNA in the appropriate puffs, strongly suggesting that the puff DNA provided the template for the synthesis of the mRNA. All together, the evidence indicated that heat had activated at least two regions of DNA, providing template for synthesis of new mRNA, which was translated to form the heat-shock proteins.

The study of heat shock in plants has not reached the high degree of sophistication described above, but it has been demonstrated that heat treatments produce heat-shock proteins in several species of plants. The technique used is the same *in vivo* labeling technique referred to previously in experiment A. Radioactive amino acid, often $^{35}$S-methionine, is given to plant cells or explant tissues. After an incubation period, the cells are broken, and debris is removed by filtration and centrifugation. The soluble proteins are subjected to electrophoresis on polyacrylamide gels in the presence of sodium dodecylsulfate (see Polyacrylamide Gel Electrophoresis in the technical insert at the end of this section). Under these conditions, the proteins separate from one another according to molecular weight. Stains may be used to show the location of high concentrations of protein, or film may be used to show the location of high concentrations of the radioactive amino acid. If the incubation period is short, the radioactive amino acid will be concentrated in those proteins synthesized most rapidly during the incubation period. Those proteins, however, are not necessarily the ones present in the highest amounts.

Figure 3–2 shows the results of an experiment using maize seedlings that were treated for twenty minutes or more at 40°C. While there was continued synthesis of most normal proteins, there was also synthesis of about ten new proteins that started within the first twenty minutes and continued for four to eight hours. Synthesis of three more proteins started one hour after the heat treatment.

Similar results are found with tobacco mesophyll protoplasts, cells released from tobacco leaves and stripped of their cell walls with enzymes. When these cells are treated at 40°C for one hour and labeled with radioactive amino acid for two hours, some proteins that are synthesized at 25°C are also synthesized at 40°C. Synthesis of some other proteins stops at 40°C. About nineteen proteins are synthesized only at 40°C. Two are synthesized transiently, whereas the rest are synthesized over a long period.

Thus, plants also make heat-shock proteins. In plants, however, heat is not so effective in turning off the synthesis of normal proteins as it is in other organisms.

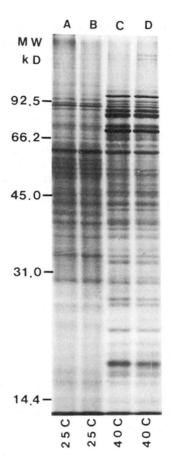

**Figure 3-2.** Patterns of protein synthesis in maize roots treated at different temperatures. A. and B. 25°C. C. and D. 40°C for 2 hours. Proteins synthesized over 2 hours were labeled with $^{35}$S-methionine. The diagram indicates positions and relative amounts of bands separated by PAGE and visualized by autoradiography. "MW" indicates the positions of standard proteins of known molecular weight. Bars at the right indicate the bands recognized as heat shock proteins. Photograph supplied by Dr. T.-H.D. Ho. (See P. Cooper and T.-H.D. Ho, *Plant Physiology* 71:215, 1983.)

What do the heat-shock proteins do? Experiments with chinese hamster fibroblasts and with *Dictyostelium* indicate that the heat-shock proteins help protect cells from damage by the heat. A mutant of *Dictyostelium* that is unable to make the smaller (26-32kD) heat-shock proteins at 40°C is quickly killed by the heat treatment. Experiments with soybeans give confirmatory results. Treatments that induce heat-shock proteins—one hour at 40°C, ten minutes at 45°C—allow seedlings to survive a subsequent two-hour treatment at 45°C. Also, a treatment with arsenite, a poison which induces proteins similar or identical to the heat-shock proteins, provides some protection against damage at 45°C.

Because heat shock involves a simple manipulation, and because there are relatively few proteins involved, it may represent a good model system for studying environmentally induced gene expression. Heat shock is of interest per se because leaves in intense sunlight may easily reach temperatures of 40°C to 45°C.

## Technical insert: PAGE (Polyacrylamide Gel Electrophoresis)

Polyacrylamide gel electrophoresis is a technique widely used to separate and identify proteins in crude or partially purified mixtures. In this technical insert we describe two common variations.

The first variation, a one-dimensional variation, can be calibrated to indicate the approximate molecular weight of a given protein.

There are three features that make this technique more powerful than earlier versions of electrophoresis. (1) The matrix. Acrylamide solutions are polymerized to form a covalently crosslinked network with spaces about the size of a protein molecule. The exact size of the spaces can be varied by changing the concentration of acrylamide (5% to 15% is usual). Proteins moving through this matrix separate according to their size, shape, and charge density. This allows a wider separation of proteins in a heterogeneous mixture. (2) The buffer system. The protein sample is placed on top of a gel that contains Cl⁻ as the principal anion. Above the protein is a buffer that has glycine as the principal anion. Glycine moves slowly in an electric field, slower than Cl⁻ and slower than the cation with which it is associated. This concentrates other anions in front of the glycine, including the protein anions. Soon after the current is turned on, the proteins form a very narrow band. This usually occurs in a region of low acrylamide concentration called the "stacking gel." The protein band then moves into the "running gel," in which the proteins separate. The narrowness of the protein band at the start is especially important for good resolution. (3) SDS (sodium dodecylsulfate). The treatment of protein samples with SDS and mercaptoethanol denatures the proteins by disrupting hydrophobic bonds and disulfide bridges, respectively. This separates polypeptide chains from one another. SDS, added to the gel buffer as well as to the sample, coats the polypeptide chains and confers a uniform negative charge density to each chain-SDS complex. Under these conditions, and in the presence of an electric field, polypeptide chains migrate away from the cathode (−) through the gel, separating mainly on the basis of molecular weight. The molecular weight of any chain can be estimated by comparing its migration rate with the migrations rates of proteins of known molecular weights (Fig. 3–3A).

After electrophoresis, proteins are fixed in place by soaking the gel in an acidic alcohol solution. The proteins can be visualized by staining with (in order of sensitivity) Amido Schwartz, Coomassie Brilliant Blue, or silver. Radioactive proteins can be visualized by autoradiography, in which a dried gel is placed close to x-ray film so that the radioactive isotope exposes the emulsion. A variant of autoradiography is fluorography, in which the gel is soaked with fluors which emit light when excited by radioactivity. The light then exposes the film. Fluorography is a way of detecting weaker radiation, since even a weak beta particle from $^3$H will excite the production of many photons.

A second variation of polyacrylamide gel electrophoresis (PAGE) increases resolution by separating proteins according to two different characteristics, isoelectric point and molecular weight. In the first stage, proteins (not treated with SDS) are subjected to one-dimensional electrophoresis in the presence of ampholytes. Ampholytes are small, synthetic polymers with a variety of acid and base groups. They, therefore, have a wide distribution of isoelectric points. Under an electric field established between a cathode (−) in a buffer of higher pH and an anode (+) in a buffer of lower pH, each ampholyte migrates until it reaches the pH at which it has a zero net charge. This is called its isoelectric point. Once there, the ampholyte serves as a buffer to maintain the pH at that point. Because there are many different

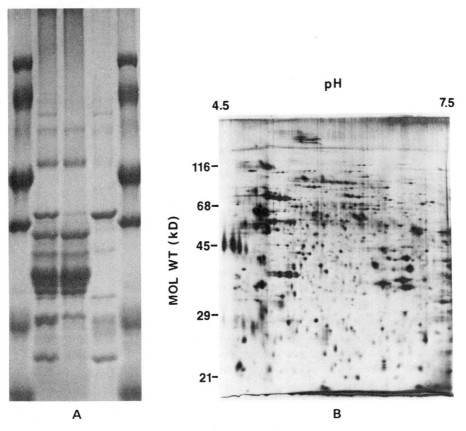

A                                                            B

**Figure 3-3.** Separation of proteins by PAGE. A. One-dimensional technique. The lanes at the left and right sides contained a mixture of standard proteins of molecular weights ranging from 14 kD (bottom) to 94 kD (top). The middle lanes contained maize thylakoid proteins. The gel was stained with Coomassie Brilliant Blue. B. Two-dimensional technique. A plasma-membrane-enriched fraction from rye epicotyls was solubilized by phenol extraction. The sample was separated first in the horizontal dimension by isoelectric point and then in the vertical dimension by molecular weight. The gel was stained with silver. Gel in A. courtesy of J. Murphy and Dr. A. Stemler. Photograph in B. courtesy of Dr. W.J. Hurkman.

ampholytes, the result is a continuous and stable pH gradient along the gel. Proteins in the sample move along this pH gradient until each reaches its isoelectric point. Since each protein accumulates at its isoelectric point, this stage is known as "isoelectric focusing." Proteins with different isoelectric points will be separated spatially along the gel.

In the second stage, the long, narrow ampholyte gel is soaked briefly in a buffer containing SDS. It is then laid along one edge of a rectangular SDS gel (corresponding to the "running gel" above). Under the electric field (with the cathode connected to the ampholyte gel edge), proteins from the ampholyte gel migrate into the SDS gel and separate according to their molecular weights. The proteins in the gel are fixed and visualized as described above. Carefully performed, this technique can resolve several hundred proteins as individual spots (Fig. 3–3B).

## B. THE PHENOLIC PATHWAY

Phenolic compounds play many roles in a plant's life. Flavonoids (flavanones and anthocyanins) give color to flowers and fruits, which presumably attract animals that aid in pollen and seed dispersal. They also are found in the epidermal cells of leaves and stems and may protect inner tissues from damage by high intensities of blue and ultraviolet light. Phenolic phytoalexins protect tissues from infection by pathogens. Phenolic compounds also form the lignin in secondary cell walls of the fibers, tracheids, and vessel elements in the xylem. Lignin is also found in the secondary cell walls of schereids in seed coats.

The production of phenolic compounds is highly regulated. This is clear from the tissue localization of the substances. There is also an additional level of control in the stimulation of synthesis by appropriate environmental influences. For instance, the production of anthocyanins in young shoots is stimulated by light (red, blue, ultraviolet, or combinations of these). The synthesis of phytoalexins is also triggered by chemical compounds that signal the presence of infectious agents.

The biochemical pathway for the synthesis of these phenolic compounds (Fig. 3–4) starts with the amino acid phenylalanine. The first enzyme in the pathway, phenylalanine ammonia lyase (PAL), converts phenylalanine to cinnamic acid. The second enzyme converts cinnamic acid to *para*-coumaric acid. At this point the pathway branches. Isozyme 1 of 4-coumarate CoA ligase converts para-coumaric acid or its relatives, sinapic and ferulic acid, to intermediates on the pathway to lignin. Isozyme 2 of 4-coumarate CoA ligase converts para-coumaric acid to 4-coumaryl CoA that goes on to become a flavonoid. Chalcone synthetase (previously called flavanone synthetase) adds three 2-carbon groups from three malonyl CoA molecules to 4-coumaryl CoA to form naringenin (flavanone). Flavanone and chalcone are isomers. Flavanone or chalcone can be oxidized, methylated, and glycosylated to form various other

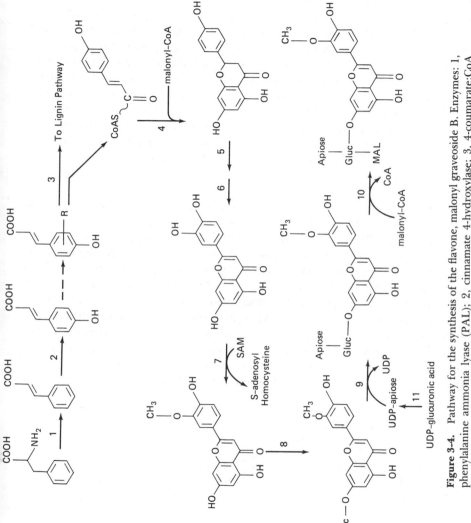

**Figure 3-4.** Pathway for the synthesis of the flavone, malonyl graveoside B. Enzymes: 1, phenylalanine ammonia lyase (PAL); 2, cinnamate 4-hydroxylase; 3, 4-coumarate:CoA ligase; 4, chalcone synthase; 11, UDP-apiose synthase. Redrawn from K. Hahlbrock, and others, *European Jour. Biochem.* 61:199, 1976.

flavonoids. The different flavonoids have different colors. In general, as flavanoids become more and more oxidized, they display a color sequence ranging from yellow to orange, red, and blue.

The induction of anthocyanin accumulation in cell cultures of parsley has been correlated with the induction of most or all of the enzymes in the phenolic pathway (Fig. 3–5). Cell cultures were chosen for study because of their convenience and homogeneity. When mature cultures, ones that have grown to maximum cell density, are diluted in fresh medium, the cells produce a burst of the first three enzymes in the phenolic pathway. These enzymes disappear after about two days. No more of the enzymes in this pathway are produced unless the cultures are illuminated. When they are illuminated, there are two new bursts of enzyme activity: one for the first three enzymes in the pathway, another (slightly delayed) one for the enzymes that are specific for flavonoids. Each burst represents the coordinated, simultaneous induction of expression of several genes. The two bursts show a coordinated, but non-simultaneous, control of expression.

Hahlbrock and his colleagues have tested whether the increase in enzyme activity for PAL and other enzymes represents synthesis of new enzyme protein. These experiments used the incorporation of radioactive amino acids as a measure of synthesis. In order to show that the label was incorporated in a specific enzyme, they produced an antibody that bound specifically to the enzyme. The antibody allowed the investigators to measure the rate of synthesis of the protein that the antibody "recognized."

To measure the rate of PAL synthesis *in vivo*, $^{35}$S-methionine was added to cells for twenty-minute intervals starting at different times after the cells had been illuminated. At the end of each twenty-minute incubation, the cells were broken and the extracted proteins were treated with antibody and other reagents to precipitate the PAL. The precipitate (anti-

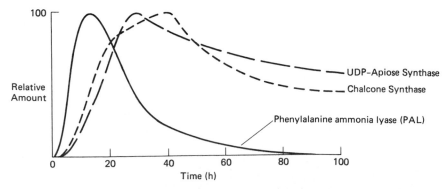

**Figure 3-5.** Induction of enzyme activites in the flavonoid pathway of cultured parsley cells. Irradiation with white light started at time 0. Redrawn from K. Hahlbrock and others, *European Jour. Biochem.* 61:199, 1976.

body plus PAL) was dissolved in SDS. The antibody and PAL were then separated by electrophoresis. Finally, the radioactivity that was associated with the PAL band was measured. Incorporation of radioactivity into the PAL band followed a pattern in time that corresponded closely to the rate of increase (and later decrease) of the enzyme activity (Fig. 3–6). This supports the idea that the increase in activity was caused by the accelerated rate of synthesis. (One should note, however, that the rate of incorporation of the labeled amino acids may have been an underestimate of the total rate of synthesis, if the labeled amino acids were slow to enter the cells.)

Does the synthesis of new PAL protein depend on the appearance of new mRNA, or does the cell start to use mRNA that is already present? To measure the amount of mRNA, the cells were extracted at different times after illumination, and RNA was purified from the crude extracts. Total RNA, both poly(A)+ RNA and polysomal RNA, were prepared. The RNAs were then tested for messenger activity in an *in vitro* protein synthesizing system by mixing them with ribosomes, amino acyl-tRNA, enzymes, and an energy source in an appropriate buffer. After an incubation period, the incorporation of $^{35}$S-methionine into antibody-precipitable protein was measured in the manner previously described. The amount of labeled protein (which measured the amount of extractable messenger activity) rose and then fell after illumination, just as the rate of synthesis of PAL measured *in vivo*. This suggests that the accelerated rate of protein synthesis depended on the increased availability of mRNA.

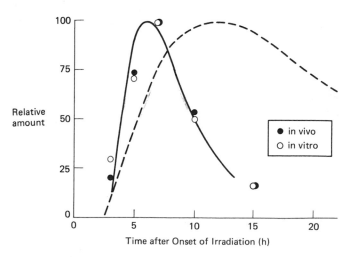

**Figure 3-6.** Induction of PAL synthesis by light. The solid line shows the level of incorporation of labeled amino acid into PAL during a 20 minute incubation. The dotted line indicates the accumulation of enzyme activity. Redrawn from K. Hahlbrock, *European Jour. Biochem.* 63:137, 1977.

Similar results were obtained with mRNA from polysomes (mRNA that is being translated) and with poly(A)+ RNA (potentially mRNA), indicating that the increased rate of synthesis represented increased availability of poly(A)+ RNA, not increased use of preexisting mRNA.

Similar results were found with other enzymes and other stimulants. The light-treated parsley cells described previously also showed increases in the synthesis of chalcone synthase and UDP-apiose synthase (an enzyme that modifies flavonoids). French bean cells treated with an elicitor from *Colletotrichum lindemuthianum,* the agent that causes anthracnose disease in beans, also synthesized PAL and chalcone synthase. The bean cells responded to infection by making a flavonoid-based phytoalexin. Parsley cells could be induced to form PAL and 4-hydroxycoumarate CoA ligase, but not chalcone synthase, by treatment with an elicitor from *Phytophthora megasperma,* another fungal pathogen. This elicitor stimulates parsley cells to form a phenolic but non-flavonoid phytoalexin.

It seems that there is a common principle here—induction of synthesis of the enzymes of the phenolic pathway. But, somehow, there is also specificity according to the species of plant and to the inducing agent. Both the mechanism of induction and the nature of the specificity are unknown.

### Technical insert:  *In Vitro* Protein Synthesis

Proteins can be synthesized by cell-free systems that contain all the components normally required by the cell for the same process. Cell-free systems were originally used to study the characteristics of the protein synthetic process itself, but now they are employed to make a particular protein from a messenger RNA. Components for the system can be prepared from tissues in the laboratory, or they can be obtained commercially. One separates the endogeneous mRNA from the other components (or treats the components with nuclease to remove endogenous mRNA) and then adds the mRNA to be tested.

Systems are commonly prepared from extracts of *E. coli* (for prokaryotic or plastid mRNAs) and from reticulocytes or wheat embryos (for eukaryotic nuclear mRNAs). For mitochondrial mRNAs, which use a somewhat different genetic code, *E. coli* components must be supplemented by mitochondrial tRNAs and enzymes.

A typical system contains the following components:

1. Ribosomes.
2. Transfer RNA.
3. Enzymes, including amino-acyl-tRNA synthetases and initiation, elongation, and termination factors (usually added as a crude mixture).
4. ATP, plus a system composed of ADP, phosphate, phosphocreatine and phosphocreatine kinase to regenerate ATP.

5. Amino acids, including at least one radioactive amino acid of high specific activity (typically, $^{35}$S-methionine or $^{3}$H-leucine).
6. Messenger RNA.
7. Salts (3 mM $MgSO_4$, 50 mM KCl) and buffer (for example, 25 mM Tris, pH 8.1).

After all the components are mixed, the solution is incubated at 37°C for between thirty minutes and two hours. After that period, incorporation of amino acids into proteins slows and stops. This is probably due to the action of nucleases on the mRNA. The synthetic processes can be measured by incorporation of radioactivity into macromolecular products. These are products that are precipitated in acid or are characterized in more specific ways, for instance, SDS-PAGE (SDS-Polyacrylamide Gel Electrophoresis).


## C. ANAEROBIOSIS AND ALCOHOL DEHYDROGENASE

Most tissues of higher plants are adapted to life in a normal atmosphere of 21% oxygen. When the supply of oxygen is drastically curtailed, they must adapt quickly to survive. This is particularly important in non-photosynthetic tissues that have no source of energy other than the catabolism of organic molecules. The study of the events that occur when a tissue becomes anaerobic helps us to understand the responses of plants to their environment. It also helps answer questions about the autogenous factors that distinguish one tissue from another since not all tissues adapt to anaerobiosis in the same way.

Anaerobiosis involves one of the most central and best-studied biochemical pathways, the glycolytic pathway, which degrades glucose to pyruvic acid. In a normal oxygen atmosphere, pyruvate enters the mitochondria and is further oxidized in a process that produces a large amount of ATP. In the oxidation process, electrons are transferred from pyruvate to oxygen. Under anaerobic conditions, pyruvate cannot be oxidized by mitochondria because there is no oxygen to accept the electrons removed in the oxidation process. In fact, under anaerobic conditions pyruvate cannot accumulate since the production of pyruvate from glucose is an oxidation and requires an electron acceptor. However, pyruvate *can* be formed if it can be reduced subsequently so that there is no net oxidation. This is what, in fact, happens. In many plant tissues, the reduction of pyruvate is a two-step process: Pyruvate is decarboxylated by the enzyme pyruvate decarboxylase to form acetaldehyde; and acetaldehyde is reduced by the enzyme alcohol dehydrogenase (using NADH as electron donor) to form ethanol. This process allows glycolysis to proceed, forming a small amount of ATP (2 ATP per glucose) as shown in Figure 3–7.

The enzyme alcohol dehydrogenase is essential to the operation of

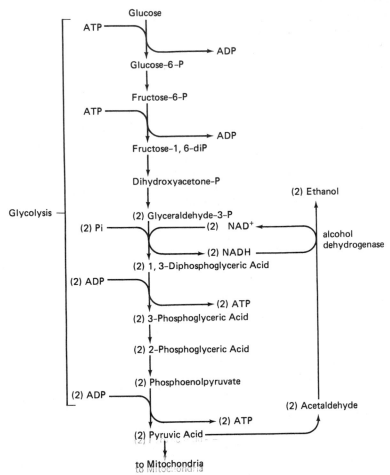

**Figure 3-7.**  Anaerobic metabolism of glucose in maize.

the glycolytic pathway under anaerobic conditions. In maize there are multiple "isozymes" of alcohol dehydrogenase formed from the subunits produced by two different alcohol dehydrogenase genes, Adh-1 and Adh-2 (Fig. 3–8). The different genes are in different locations in the maize genome. Each gene produces a polypeptide chain, which is an enzyme subunit. Two chains can join together to form the completed alcohol dehydrogenase enzyme. Since the polypeptide chains may differ in electric charge, the enzyme molecules formed from different combinations of subunits may be distinguished on the basis of their electrophoretic mobility in a starch gel medium.

When maize tissues are placed in a nitrogen atmosphere, the synthesis of most proteins stops. This is explained by the loss of energy supply,

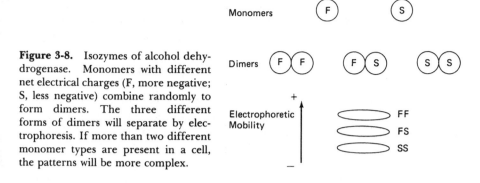

**Figure 3-8.** Isozymes of alcohol dehydrogenase. Monomers with different net electrical charges (F, more negative; S, less negative) combine randomly to form dimers. The three different forms of dimers will separate by electrophoresis. If more than two different monomer types are present in a cell, the patterns will be more complex.

since a considerable amount of ATP or its equivalent is needed for protein synthesis. At the same time, however, there are about twenty proteins (ten major and ten minor), whose synthesis remains at the same level or increases. One of these proteins is alcohol dehydrogenase. The presence of this enzyme apparently helps the tissues survive in the anaerobic conditions, since mutants that do not form this enzyme die faster in water-saturated, anaerobic soils. Three other proteins are aldolase, phosphoglucoisomerase, and pyruvate decarboxylase. All are enzymes of the anaerobic glycolytic pathway. The identity and functions of the other proteins produced under anaerobiosis are not known, though it is clear they are not heat-shock proteins. (It must be recalled that in *Drosophila* anaerobic conditions did induce the synthesis of certain of the heat-shock proteins.)

The expression of the Adh genes is tissue specific. In the scutellum (the maize seed cotyledon), only the Adh-1 gene is induced; no Adh-2 gene product appears. Not only that, but each gene may have different alleles. These alleles make subunits that can be distinguished by electrophoretic mobility and that may also be expressed in a tissue-specific way. The S ("slow," for slow electrophoretic mobility) allele of Adh-1 is induced in the scutellum to a slightly higher degree than is the F ("fast") allele. This is in contrast to the situation in roots, where the F allele is induced to a much higher degree than the S allele. Thus the alleles of Adh-1 contain some information that quantitatively governs their expression in different tissues, in addition to the information that governs the primary structure (F and S electrophoretic mobilities) of the subunits. The two types of information are tightly linked, since in controlled matings no recombination between tissue-specific expression and mobility is observed. However, the two types of information are not identical, since it has been possible to convert an S allele to an F allele with a mutagen, ethylmethane sulfonate, without changing the way that allele is expressed in scutellum and roots.

The mechanism by which the plant induces the expression of the Adh genes under anaerobic conditions remains a major unanswered question; the reason for the tissue specificity of Adh induction is another mystery.

The mechanism of the induction has been tested by assaying for the Adh mRNA. Does the mRNA appear only under anaerobic conditions, or is it already present, to be used when the oxygen supply gets low? The question is the same as that asked previously about the phenolic pathway. But the technique to assay mRNA used by Freeling, Taylor, and their colleagues at the University of California at Berkeley and by Peacock's laboratory in Australia was quite different from the technique discussed in regard to the phenolic pathway. Rather than assaying ability to direct protein synthesis *in vitro*, the workers at these laboratories synthesized DNA complementary to the mRNA for alcohol dehydrogenase (cDNA). They then made the cDNA radioactive ("labeled" the cDNA) and tested for mRNA by its ability to hybridize to the labeled cDNA.

The preparation of cDNA is a multistep process. Often, as in this case, it involves not only the initial synthesis of cDNA but also the construction of bacterial clones that contain and produce individual base sequences of cDNA and the selection of those clones that produce just the sequence that one wants. The cloning steps are important because in principle they provide an unlimited supply of a homogeneous DNA base sequence.

Some technical aspects of cDNA synthesis, its cloning and use, are detailed in the inserts following this section. Here we describe how these techniques were applied to Adh (Fig. 3–9). The first step was the synthesis of cDNA using mRNA as a template. The mRNA should contain the highest possible percentage of Adh-specific mRNA. Peacock and his collaborators subjected maize roots to anaerobic conditions for twenty-four hours and extracted polysomes from these roots. The mRNA was released from the polysomes with detergent and EDTA (which chelates the $Mg^{2+}$ needed to hold ribosomes together) and was purified by binding it to and eluting it from a column containing oligo-dT attached to an inert, stationary material. The oligo-dT, a short segment of single-stranded poly(T) DNA, binds to poly-A regions so that this step selected poly(A)+ RNA, virtually all of which was mRNA. The poly(A)+ RNA was sedimented through a sucrose gradient in an ultracentrifuge. The investigators collected the size class that would serve as a template for 40kD proteins, the size of the Adh subunit. This partially purified fraction of RNA was then combined with the enzyme "reverse transcriptase" and the nucleoside triphosphate substrates. The enzyme used the RNA as a template and formed cDNA. Another step made the cDNA double stranded. The result was a population of cDNA molecules that reflected the population of mRNA templates.

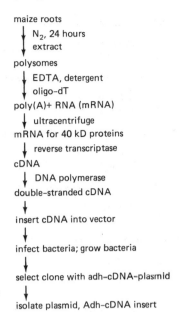

maize roots

↓ N$_2$, 24 hours
↓ extract

polysomes

↓ EDTA, detergent
↓ oligo-dT

poly(A)+ RNA (mRNA)

↓ ultracentrifuge

mRNA for 40 kD proteins

↓ reverse transcriptase

cDNA

↓ DNA polymerase

double-stranded cDNA

↓

insert cDNA into vector

↓

infect bacteria; grow bacteria

↓

select clone with adh–cDNA–plasmid

↓

isolate plasmid, Adh–cDNA insert

**Figure 3-9.** Strategy for making alcohol dehydrogenase (Adh) cDNA.

The second step was the formation of bacterial clones containing cDNA. The strategy for inserting DNA into bacteria involves plasmids, small circular DNA molecules. These are infectious, in that they can enter bacterial cells with high efficiency. Once in a bacterial cell, a plasmid can be replicated many times outside the main bacterial chromosome. Or it can be incorporated into the chromosome and replicated with that molecule. Plasmids serve as "vectors" for inserting other sequences of DNA, because even when other sequences are added to it the plasmid can enter a bacterium and be replicated. In order to insert other sequences in the plasmid circle, one must cut the circle. This is done with a restriction enzyme, an enzyme that hydrolyzes DNA at a specific base sequence. The many different restriction enzymes can be used to cut at different sequences (Fig. 3–10). We can choose a plasmid and a restriction enzyme that makes a single cut in the plasmid so that we open the circle of DNA in a specific place. We can insert cDNA and reclose the circle by forming "sticky ends," single-stranded DNA with complementary sequences, on the cDNA and the open plasmid. Peacock's colleagues added a string of C bases to the cDNA and a string of G bases to the open plasmid. When mixed together in warm solution and slowly cooled, the C tails and the G tails paired, and the result was a new circle made from one cDNA and one plasmid. *E. coli* were infected with the plasmids. Only hybrid plasmids, the ones with cDNA inserts, entered the cells because only they formed circles and only circles are infectious.

The final step was the selection of those bacteria that had the desired cDNA sequences. Bacteria were "cloned" by growing them on agar at a

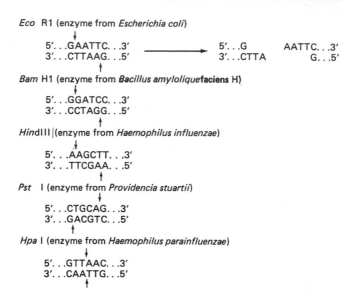

**Figure 3-10.** Some examples of restriction enzymes as defined by the sequences of DNA that they cut. Dots indicate unspecified bases in the DNA chains. Arrows show the cutting points. The example at the top shows how the chains look as they separate. Note that each sequence is a palindrome; that is, it has a 2-fold rotational axis of symmetry.

concentration so dilute that individual cells produced individual colonies (clones). Then samples from each clone were tested individually for the presence of the DNA sequence desired. There are many ingenious ways of doing this efficiently. Peacock's colleagues transferred cells from each clone to two filters by pressing the filters on the Petri dish containing the colonies. The cells on the filters were grown into new colonies. The colonies were lysed with NaOH under conditions in which the DNA that emerged stuck to the filters. One filter was then incubated with radioactively labeled cDNA from anaerobic roots and the other, with cDNA from aerobic roots. Of 4,505 clones tested, 105 contained DNA that was complementary to cDNA from anaerobic roots but *not* aerobic roots. The 105 were examined individually by testing whether or not their DNA would hybridize to mRNA that would serve as a template for alcohol dehydrogenase in an *in vitro* protein synthesizing system. Twelve of the 105 clones passed this test. DNA from the two clones with the longest cDNA inserts served as a source of cDNA. These DNAs, unlike those produced directly from a population of mRNA templates, represented homogeneous (pure) DNA. Within the limits of the selection methods used to obtain the clones, we can be sure that these DNAs had the base sequence of the Adh gene.

The cDNA can be made radioactive by incubating it with *E. coli* DNA polymerase I and $^{32}$P-labeled deoxyribonucleoside triphosphates. DNA polymerase I binds to "nicks," that is, single-stranded breaks, in the

DNA. At each nick there is a free 3′ and a free 5′ end of the broken polynucleotide chain. Polymerase I removes nucleotides from the 5′ end, but at the same time it adds nucleotides (from the $^{32}$P-deoxyribonucleoside triphosphates) to the 3′ end, using the unbroken strand as a template. The result is a new, radioactive strand of DNA with the same base sequence as the old strand. This process is known as nick translation, because the nick moves along the DNA as the enzyme works.

The labeled cDNA can be used to test for the presence of Adh mRNA in aerobic and anaerobic tissues by a technique known as "blot hybridization" or "Northern blots." In this technique RNA is extracted from tissues, purified to remove DNA and protein (and sometimes to select the poly(A)+ fraction), and subjected to electrophoresis in agarose gels. This separates the RNA into size classes. RNA in the gels is then transferred to filter paper by placing the gel on the filter and drying. In this process the RNA sticks irreversibly to the filter. The filter is incubated with a solution of the labeled cDNA (the "probe") under conditions in which the cDNA can hybridize to the RNA where the RNA is present. The unhybridized cDNA is washed off, and the cDNA on the RNA that is stuck to the filter is detected by autoradiography.

Using this test, Peacock's group showed that the amount of Adh-1 mRNA in anaerobic maize roots was about 50 times greater than that in aerobic roots (Fig. 3–11). When aerobic roots were placed under anaerobic conditions, the amount of RNA started to increase in two hours and reached its maximum by five hours. When anaerobic roots were returned to aerobic conditions, the Adh mRNA did not degrade rapidly—the half-life was greater than eighteen hours—suggesting that the control involved

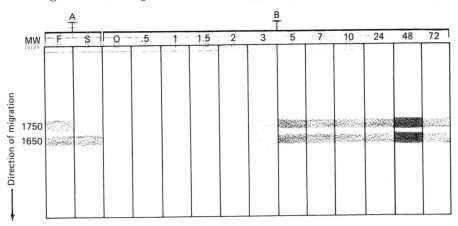

**Figure 3-11.** Assay of Adh mRNA extracted from anaerobic roots of maize. A. Different maize varieties. Left: Adh-1F homozygote. Right: Adh-1S homozygote. B. Induction by anaerobiosis. From left to right: 0, 0.5, 1, 1.5, 2, 3, 5, 7, 10, 24, 48, 72 hours of anaerobiosis. Photograph kindly provided by W.L. Gerlach. See Gerlach and others, *Proc. Natl. Acad. Sci.* USA 79:2981, 1982.

its synthesis rather than its decay. Although in this study it was not determined if the appearance of mRNA involved increased supply or processing of nRNA, the method could, in principle, be used to detect nRNA as well as mRNA and thus to answer this question.

Another interesting finding was that the Adh-1 F allele produced two distinct types of mRNA differing in size (Fig. 3–11). One was about 1,650 nucleotides long; the other, about 1,750. The Adh-1 S allele, on the other hand, produced only one mRNA which was 1,650 nucleotides long. The significance of this difference is not clear, but it seems possible that the Adh-1 F allele has two promoters controlling RNA synthesis. If these promoters functioned differently in scutellum and roots, they could help explain the tissue-specific expression of the F and S alleles.

In summary, anaerobic conditions lead to the induction of specific genes that help plant tissues survive without oxygen. The induction of alcohol dehydrogenase involves the accumulation of new mRNA. Further work is necessary to determine the mechanism by which the level of mRNA is controlled and to determine why different alleles are expressed preferentially in different tissues.

### Technical insert: cDNA (Complementary DNA, Copy DNA)

cDNA is formed from mRNA. Sometimes it is made using a purified mRNA, but usually it is made from a mixture of mRNAs. In the latter case, the cDNA will be a mixture too, until it is cloned or otherwise purified. The steps in making cDNA (in addition, see Fig. 3–12) are as follows:

1. Obtain mRNA. Poly(A)+ mRNA is especially useful for this because the poly(A) tail binds to oligo-dT DNA (single-stranded poly(T) DNA, which is commercially available). The oligo-dT serves as a primer in the first DNA synthesis reaction (see next step).

2. Copy mRNA with the enzyme reverse transcriptase (RNA-dependent DNA polymerase). This enzyme is obtained from cells infected with an RNA virus that makes a DNA intermediate for the purposes of replication (for example, chicken cells infected with avian myeloblastosis virus). The reaction requires deoxyribonucleoside triphosphates as substrates, as well as a template (mRNA) and primer (DNA bound to the mRNA). Oligo-dT is a good primer for use with poly(A)+ RNA. The product of the reaction is true cDNA.

3. Degrade mRNA with base.

4. Make double-stranded cDNA using the cDNA as template, DNA polymerase I from *E. coli,* and deoxyribonucleoside triphosphates as substrates. A hairpin loop at the end of the cDNA, often formed by reverse transcriptase, may serve as primer.

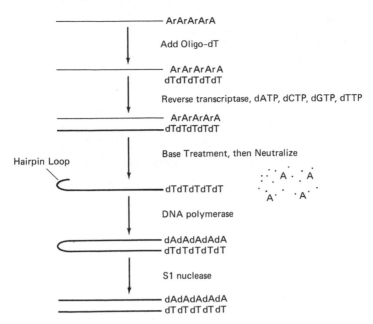

**Figure 3-12.**  Preparation of cDNA.  The light line represents RNA.  The heavy line represents DNA.

5. Treat with S1 nuclease (an enzyme that degrades only single-stranded nucleic acids) to remove the hairpin loop.

The double-stranded cDNA (abbreviated as "cDNA" in the text and in most other descriptions of the subject) may be used as a probe in hybridization assays. It may also be used for cloning.

### Technical insert:  Cloning

To clone DNA usually means to induce a bacterium to replicate a particular sequence of DNA in a form that allows its easy recovery and separation from the rest of the DNA in the bacterium. The technique involves infecting the bacterium with a plasmid or virus (a "vector") that has been linked to the desired DNA sequence (Fig. 3–13). Vectors are DNA molecules that can enter the bacterial cells and then have the ability to replicate themselves autonomously, that is, separated from the chromosome of the cell. The bacterium produces vector as it reproduces itself. The vector can then be extracted from the cells and purified, and the desired sequence can be removed and separated from the unwanted part of the vector.

The choice of vector is important. A common plasmid used to clone cDNA is pBR322. This plasmid's small size allows it to be easily purified. It

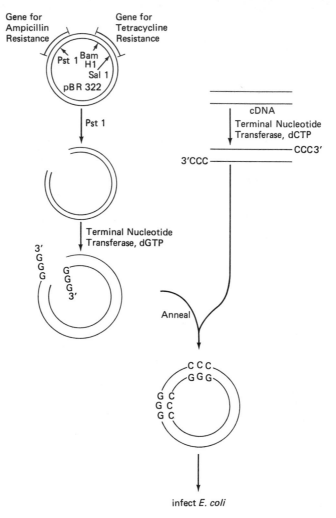

**Figure 3-13.** Cloning of cDNA in pBR322, a plasmid containing genes for re-sistance to two antibiotics. Arrows in the center of the plasmid circle (top left) show the only sites at which the plasmid is cut by the restriction enzymes, *Pst* I, *Bam* HI, and *Sal* I.

also contains two special genes. One gene, for ampicillin resistance, can be cut with the restriction enzyme *Pst* I; another gene, for tetracycline resis-tance, can be cut with either of two restriction enzymes, *Bam* H I or *Sal* I. Then, cDNA can be inserted in a cut when the circle is reformed. If it is in-serted then, the gene into which it is inserted will not be active. Thus one can detect plasmids with inserts because the bacteria that contain the inserts do not acquire antibiotic resistance.

A typical protocol for cloning cDNA is as follows:

1. Treat pBR322 with Pst I.
2. Form poly-dG at the 3' ends of the linearized pBR322, using the enzyme terminal transferase and GTP as the substrate. Repurify the DNA.
3. Form poly-dC at the 3' ends of the cDNAs, using terminal transferase and CTP. Repurify the DNA.
4. Combine the G-tailed pBR322 and C-tailed cDNAs and anneal (join) the two by heating to 63°C and then slowly cooling. The connections can be made permanent (covalent) using DNA polymerase and DNA ligase, but this is not necessary for the next step. The recipient cell apparently repairs the DNA itself.
5. Infect a strain of *E. coli* that is ampicillin sensitive and tetracycline sensitive with annealed plasmids. Plate a dilute suspension on agar, so that colonies represent individual bacteria. Select colonies that are ampicillin sensitive but tetracycline resistant. These have plasmids that include cDNA inserts.
6. Grow the bacteria. Lyse and purify the plasmid DNA by CsCl centrifugation in the presence of ethidium bromide. CsCl centrifugation separates molecules by density. Ethidium bromide allows one to separate small circular (plasmid) DNA from large circles (chromosomes) and from linear pieces (broken chromosomes). Helical stress prevents the small circles from absorbing as much ethidium bromide as the large circles and the linear pieces. As a result, the small circles stay denser and migrate farther into the CsCl. Treat plasmid with *Pst* I to separate the cDNA insert from the rest of the plasmid. Separate the fragments by electrophoresis.

## Technical insert: Blot Hybridization

Blot hybridization is a technique for transferring and hybridizing specific sequences of nucleic acids that have been separated by gel electrophoresis. Because the original report dealing with transfer of DNA pieces was authored by Edward M. Southern of Edinburgh, any application of his work to DNA is called a "Southern blot." Since DNA transfer was a "Southern blot," it seemed natural to give the nickname "Northern blot" to the transfer of RNA. Extending the geographical whimsey, transfer and detection of proteins separated by electrophoresis are often called "Western blots."

The procedure for Southern blots is as follows:

1. Fractionate the DNA mixture by electrophoresis on agarose gels. Agarose is a polysaccharide. In the form of a gel, the polysaccharide molecules form pores that resist the movement of DNA pieces. Under an electric field, DNA pieces move through the pores at rates proportional to their lengths.

2. Treat the gel to denature the DNA (for example, by soaking in base). Return to nondenaturing conditions (soak in neutral buffer).

3. Cover the gel with a nitrocellulose filter and then put paper towels on the filter, weighing them down to assure even contact. The towels pull solution through the filter by capillary action. The DNA sticks to the filter, forming "blots." The DNA is fixed to the filter by heating it at 65°C to 80°C for about two hours.

4. Incubate the filter with a solution (0.1 ml per square cm of filter) containing labeled probe DNA that has been denatured. When the probe DNA is denatured, the strands are separated and can hybridize to complementary DNA on the filter. The precise conditions of temperature and salt concentration are chosen to allow hybridization with a favorable balance between high sensitivity and low "noise" (nonspecific hybridization).

5. Wash with buffer to remove the non-hybridized probe. Temperature and salt concentration can be varied to optimize detection and background.

6. Visualize the location of the probe and find the location of DNA complementary to the probe by autoradiography or fluorography.

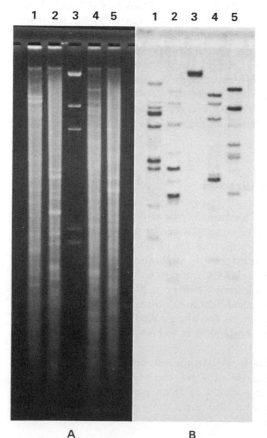

**Figure 3-14.** Southern blot hybridization. A. Total DNA from pea (lanes 1 and 2), *lamdba* bacteriophage (lane 3), and mung bean (lanes 4 and 5) separated by agarose gel electrophoresis and stained with ethidium bromide. Note the smears of stain in the lanes containing plant DNA, indicating the presence of DNA pieces of many different sizes. B. The same DNA, transferred to a nitrocellulose filter, hybridized to a $^{32}$P-labeled, cloned DNA probe, washed, and used to expose autoradiographic film. The bands show the sizes of DNA fragments complementary to the probe. Photograph courtesy of Lon Kaufman and Jon Watson.

A                                              B

Note that each blot gives three pieces of information. The number of different types of DNA that are complementary to the probe is shown by the number of bands, assuming that each different type is a different size. The sizes of those molecules of DNA can be calculated, if the gel has been calibrated with standard sized pieces of DNA. The amounts present are demonstrated if the staining intensity has been calibrated with standard amounts of DNA. (See Fig. 3–14.)

## D. AUXIN-INDUCED GENE PRODUCTS

Auxin has been a key factor in many classical studies of plant development. Many processes depend on its presence: cell elongation, phototropism and gravitropism, xylem differentiation, apical dominance. We know quite a bit about how auxin is formed, about how it is metabolized, and even a little about how it moves from cell to cell. We can describe what happens when auxin reaches the cell, but how auxin works remains, largely, a mystery. In particular, why auxin has different effects on different cells is totally unknown.

The idea that auxin controls gene expression has a long history. As early as 1953, Skoog showed that auxin stimulated RNA synthesis in tobacco cultures. Key and Shannon demonstrated the same effect in soybean hypocotyls. Actinomycin D, an inhibitor of RNA synthesis, and cycloheximide, an inhibitor of protein synthesis, both prevented auxin's stimulation of growth in coleoptiles and pea stems. However, the extra RNA that was produced in the presence of auxin in these experiments turned out to be primarily ribosomal RNA. It was impossible, using the data available up to 1970, to conclude that auxin stimulated the expression of any genes with specific developmental functions.

At the same time, physiological studies of auxin action tended to minimize the role of selective gene expression. One basic effect of auxin is a stimulation of cell enlargement, as observed in oat coleoptiles, pea epicotyls, and soybean hypocotyls. These tissues respond to treatment with auxin rapidly, often within ten minutes. The speed of this response was cited in arguments that auxin does not work by inducing the expression (transcription) of specific genes—it was thought that such induction takes longer than ten minutes in plant cells. Another factor was the acid-growth hypothesis, which asserts that auxin acts by stimulating the acidification of the cell-wall space. This hypothesis does not invoke a direct stimulation of RNA or protein synthesis by auxin. Evidence in support of the acid-growth hypothesis, including the observation that aspects of auxin-induced growth could be mimicked by low-pH buffers, tended to focus attention away from gene expression.

However, Larry Vanderhoef and his colleagues pointed out that there are two stages to auxin-induced growth (Fig. 3-15). For instance, a

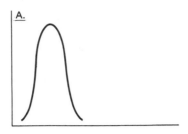

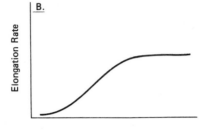

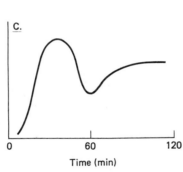

Elongation Rate

0                60               120

Time (min)

**Figure 3-15.** Stages in the stimulation of growth by auxin, according to the observations of Vanderhoef and his associates. A. Rapid stimulation in elongation rate mediated by acidification of the cell wall. B. Slower, long-lasting, auxin-specific stimulation, not mimicked by acid. C. Kinetics of stimulation expected from the addition of processes in A. and B. Patterns like these have been observed in auxin-treated soybean stem segments and some other tissues.

report in 1981 by Vanderhoef and Dute on the kinetics of auxin-induced elongation of soybean and pea stem segments demonstrated an early phase of elongation that started at ten minutes and was mimicked by low-pH buffer. A later phase started at about one to two hours after auxin treatment and was not induced by buffer. The existence of this second stage gave new impetus to the study of gene expression in those tissues that show auxin-induced growth.

Rapid auxin-induced changes in the rates of synthesis of specific proteins, and in the amounts of mRNA molecules coding for their synthesis, have now been demonstrated using many of the techniques we described earlier in this chapter. One such demonstration, by Theologis and Ray at Stanford, used etiolated pea seedlings. Stem sections were cut and incubated for two hours to deplete them of auxin. Then, they were incubated with or without indole acetic acid (IAA) for periods ranging from ten minutes to two hours. Poly(A)+ RNA was isolated from the sections and added to a wheat germ *in vitro* protein synthesizing system with $^{35}$S-methionine as a radioactive label to allow the detection of small amounts of protein. The proteins synthesized *in vitro* were separated by two-dimensional SDS-PAGE, and the proteins were visualized by autoradiography. The autoradiograms representing sections incubated with and without auxin both had many spots, just as one would expect, since even without auxin the tissue was young and healthy and actively making proteins. However, there were at least five spots that appeared or increased after the addition of auxin, and three of these showed initial increases after ten to twenty minutes of treatment. This is strong evidence that auxin increases the activity of a few specific mRNA species.

A second study, by Walker and Key, prepared clones of cDNA specific for auxin-induced mRNA species. Elongating sections were cut from the hypocotyls of dark-grown soybean seedlings. Poly(A)+ RNA was extracted and purified from these sections, and it was used as a template for the synthesis of cDNAs. The cDNAs were inserted into the ampicillin gene of pBR322 plasmids and were used to transform *E. coli,* in the same manner as was done in the Adh work. This situation was different from the Adh case, however, because auxin, unlike anaerobiosis, did not reduce the level of the majority of the mRNAs that were normally expressed. This meant that the mRNA for auxin-induced genes was a much lower fraction of the total, and more clones, 14,000, had to be tested. Nevertheless, the initial screening was similar. Colonies were grown on filters and lysed so that their DNA stuck to the filters; then the filters were incubated with labeled cDNAs prepared from poly(A)+ RNAs from auxin-incubated or auxin-depleted hypocotyls. Of the 14,000 clones, nineteen were selected as showing hybridization to auxin-incubated, but not auxin-depleted, cDNA. Two of these were finally used.

The cDNAs, labeled with radioactive nucleotides, were used to detect poly(A)+ RNA that had been subjected to electrophoresis and then transferred to filters (Northern blots). The electrophoresis established the size of the mRNA. The amount of label that stuck to the filter was a measure of the mRNA that was present. The two clones detected different mRNAs of slightly different sizes. The mRNA complementary to clone 1 was present in higher amounts than the mRNA complementary to clone 2.

In control tissues, the mRNAs were present in higher concentrations in the elongating zones than in the zones above or below the elongating region. In the elongating zones, the mRNA complementary to clone 1 maintained approximately the same concentration if the hypocotyls were incubated in a solution of the auxin 2,4-D, but decreased with a half-life of one to two hours if the hypocotyls were incubated without auxin. The

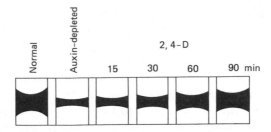

**Figure 3-16.** Northern blot hybridization showing the induction of a poly(A)+ RNA by auxin. RNA was prepared from (left to right) normal tissue, auxin-depleted tissue, auxin-depleted tissue incubated with 2,4-D for 15, 30, 60, and 90 minutes. Bands were visualized by autoradiography after transfer to a nitrocellulose filter and hybridization to a cloned, radioactive probe. Redrawn from data in Walker and Key, *Proc. Natl. Acad. Sci.* USA 79:7185, 1982.

mRNA complementary to clone 2 increased if the tissues were incubated in 2,4-D and decreased in the absence of auxin, reaching an undetectable level in about six hours. The mRNA levels climbed rapidly when auxin-depleted tissues were treated with auxin. The rise of mRNA in clone 2 was apparent as soon as fifteen minutes after auxin was added (Fig. 3-16).

Theologis and his colleagues have used cDNA clones to test whether the rapid appearance of auxin-specific RNAs depends on the $H^+$ secretion that occurs soon after auxin is added. Neither fusicoccin, which stimulates $H^+$ secretion, nor vanadate, which inhibits it, had an effect on the induction of the RNAs. This suggests that the induction of the RNAs is more directly related to the presence of auxin than to ion transport or growth.

All these results are consistent with the hypothesis that the two clones of cDNA detect RNAs that have something to do with elongation. It is clear that auxin can have very rapid effects on gene expression, and thus the speed of auxin action cannot be used to dispute the possibility of control at this level. Certainly, the effect of auxin is rapid enough to account for the second phase in elongation that was observed by Vanderhoef and Dute. However, the mechanism by which auxin increases the amount of RNA is not defined. How auxin works remains a mystery, but we know more than we did. And we now have tools to study more intimate details of auxin action.

Auxin has many other effects apart from stimulating elongation. For instance, it induces xylem differentiation, adventitious root formation, and callus in wounded tissues. It will be interesting to compare the effects of auxin on gene expression in these systems. Some experiments by Zurfluh and Guilfoyle on soybean hypocotyls already indicate that more and different gene products appear in mature tissues treated with auxin than in the immature, elongating tissues.

## E. SEED PROTEIN

Storage proteins are accumulated in the cells of a seed's endosperm or cotyledons; they serve as a storage form of nitrogen, carbon, and energy that can be released to nourish the growing embryo when the seed germinates.

The formation of seed storage proteins presents a marvelously appropriate system for analysis by the techniques of molecular biology. When a seed is formed, there is a period when these proteins are synthesized in large amounts. At this time, their mRNAs represent a large fraction of the functional mRNA in the cells where they accumulate, similar to the situation in reticulocytes, which dedicate their protein synthetic machinery to making hemoglobin. This makes it relatively easy to make and clone cDNAs for the seed protein genes. The system of protein storage in seeds is valuable, too, because the control of the expression of the seed protein genes is interestingly complex, involving coordination with other events involved in seed and fruit formation and the coordination between different seed protein genes. Also involved in the process is tissue specificity, organelle specificity, post-transcriptional modification of RNA, and post-translational modification of protein. Finally, the seed storage proteins of certain plants are especially important, because they form the basis of nutrition of humans and other animals.

Among the cereals, maize has been studied most extensively. Seed proteins are found in membrane-bound protein bodies within the cells of the endosperm (Fig. 3–17). The protein bodies are formed from cisternae of the endoplasmic reticulum (ER). As proteins are synthesized by ribosomes on the surface of the ER, they move across the ER membrane and precipitate within the lumen. Aggregations of the precipitate, together with the surrounding membrane, squeeze off and eventually detach from the ER.

The maize seed proteins are called "zein," and there are four classes of them, differing in molecular weight as determined by SDS-PAGE. The 22kD and 19kD classes are heterogeneous with respect to isoelectric point (Fig. 3–18). This heterogeneity suggests that there may be several genes that produce proteins with related amino acid sequences. The existence of several genes has been confirmed directly by showing that the sequences in the genome that hybridize to a seed protein cDNA are found in multiple copies in more than one part of the genome.

A model of the 19kD zein protein, including its amino acid sequence, has been described by Larkins on the basis of the base sequence of a cloned zein cDNA. The clone was produced in the normal way, using mRNA of membrane-bound polysomes extracted from developing corn seeds. The cDNA was formed, inserted into a pBR322 plasmid, and used to transform *E. coli*. Samples of clones were selected, and their plasmids

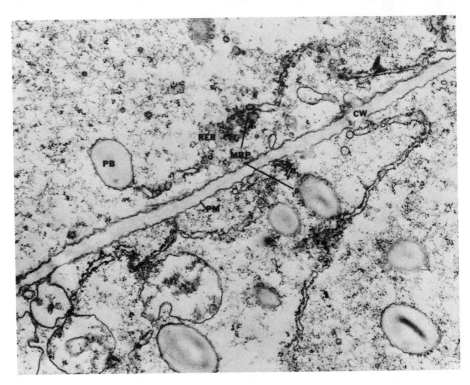

**Figure 3-17.** Electron micrograph of a developing maize endosperm cell, showing protein bodies (PB) budding from rough endoplasmic reticulum (RER). MBP, membrane-bound polyribosomes; CW, cell wall; PM, plasma membrane. Reprinted from B.A. Larkins and W.J. Hurkman, *Plant Physiology* 62:256, 1978. Photograph supplied by Dr. Larkins.

demonstrated the ability to hybridize to a major class of mRNA and to hybridize to an mRNA that codes for the synthesis of zein proteins *in vitro*.

The base sequence of one cloned cDNA, corresponding to a 19kD zein protein, was determined by a simple procedure that creates DNA pieces of sizes that reflect their base sequence. The procedure begins with a homogeneous DNA template. One may use the cDNA removed from a cloned plasmid with a restriction enzyme. If the cDNA is too large (over 300 base pairs), it can be broken into smaller pieces with other restriction enzymes and the pieces purified by electrophoresis. One end of one complementary strand is labeled specifically with a radioactive nucleotide, so that every piece that includes that end can be visualized by autoradiography. Then four samples are treated in four different ways so as to break the DNA specifically (or preferentially) at the A, G, C, or T bases. The treatment must be inefficient, so that it forms a population of DNA

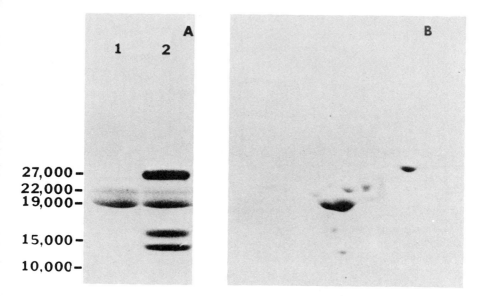

**Figure 3-18.** One dimensional (A) and two dimensional (B) PAGE of zein proteins. In B, separation by isoelectric point is from left to right; separation by molecular weight is from bottom to top. Reprinted from B.A. Larkins, L. Pedersen, M.D. Marks, and D.R. Wilson, *Trends in Biochemical Science* 9:306, 1984. Photograph supplied by Dr. Larkins.

pieces, some of which have been broken at each base of the correct type. Those pieces will differ in size, according to the positions of the bases that were broken. Each sample is subjected to electrophoresis in polyacrylamide gel to separate the DNA pieces by size. The four samples corresponding to the four bases are run side by side, so that they can be compared. From the autoradiograms one can read the sequence directly. (See the technical insert at the end of this section.)

When the zein cDNA base sequence was translated into an amino acid sequence using the genetic code, a surprising degree of regularity appeared (Fig. 3–19). The central portion of the protein was made up of nine copies of a highly conserved sequence of twenty amino acids. The nine sequences were in tandem, that is, connected end-to-end, with four glutamines at each connection. Additional sequences were placed at the N-terminal and C-terminal ends of the protein. There was a "signal sequence" that was synthesized at the N-terminal end, but it was removed as the protein was transported across the membrane of the ER. Larkins has postulated that the repeated sequences form nine alpha helices arranged in a large cylinder with the terminal sequences at each end. Many cylinders would join together in a regular array to form the solid mass that fills protein bodies (Fig. 3–20).

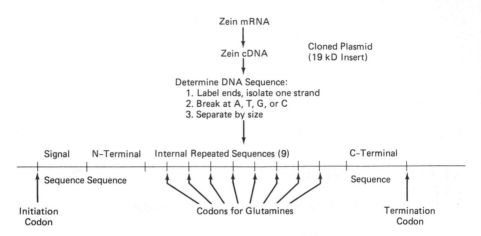

**Figure 3-19.** A zein cDNA. From the cDNA base sequence, one can infer the amino acid sequence of the protein.

Seed proteins of legumes have also been studied thoroughly. These are found in the cells of the cotyledons. They are synthesized and then transported to the vacuole, which becomes packed with protein precipitate and eventually becomes the organelle we recognize as a protein body.

There are three size classes of proteins in soybean protein bodies. The smallest proteins are designated 2S (S represents the rate of sedimentation of the proteins in an ultracentrifuge). They are enzymes and proteinase inhibitors. The middle-sized proteins, 7S in size, are called "conglycinin" and can be further subdivided into beta and gamma types. Beta-conglycinin has four subunits, though only three at a time join together to form any one molecule. The largest proteins, 11S in size, are called "glycinin." Each glycinin molecule has three 40kD subunits and three 20kD subunits. It is thought that the glycinin molecule is formed by the association of three precursors of 60kD or larger that are cleaved to yield the three 40kD and the three 20kD subunits. There is considerable heterogeneity within the subunits of conglycinin and glycinin. As in maize, this reflects the presence in the genome of many genes for these proteins.

As in maize, legume seed proteins are synthesized on the rough ER. As they move across the ER membrane, they lose their signal sequences. At approximately the same time, some of the proteins (7S storage proteins and lectins, but not 11S storage proteins) gain oligosaccharide chains. The oligosaccharides, composed of three glucose, nine mannose, and two glucosamine residues, are attached to asparagine side chains. From the ER, the proteins move to the Golgi complex, where at least some of the oligosaccharides are modified. In the Golgi complex, glucose

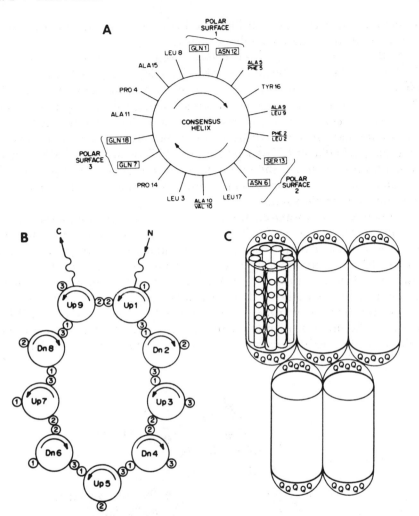

**Figure 3-20.**  Model for the three-dimensional structure of zein proteins and the aggregation of proteins within a protein body. Each protein has nine alpha-helical segments, connected by glutamines (Q). A. Positions of the side chains in the alpha-helices as seen from one end. B. Positions of the nine alpha-helices as seen from one end. C. Model of five proteins in the aggregated state. Reprinted from P. Argos, and others, *Jour. Biological Chemistry* 257:9984, 1982.

and mannose residues are removed from the proteins, and various sugars may be added. Then the proteins travel by way of dense Golgi vesicles to the vacuole. In the vacuole, there is further processing of the proteins by proteolytic cleavage and modification of the oligosaccharides.

Several research groups have studied the seed protein genes, focusing on the mechanism by which gene expression is turned on during seed development. In each case, an increase in the level of mRNA present in the cells seems to be the controlling factor.

Goldberg's group has used cDNA clones to study the presence of mRNA molecules for soybean seed proteins (Fig. 3–21). A first test of developing soybean seeds revealed that mRNAs complementary to the cDNAs were present in only low amounts during the early phases of embryogenesis. These mRNAs increased over 100-fold in the midmaturation phase, when seed protein synthesis was rapid. In midmaturation cotyledons, four particular mRNAs represented 25% of the total mRNA. After the midmaturation stage, the amounts of these mRNAs dropped, so that by the time the seeds went dormant the levels were nearly the same as before the midmaturation stage.

A second test compared the amount of mRNAs complementary to the cloned cDNAs in leaves with the amount in midmaturation embryos. It was found that leaf nuclear RNA contained less, quite possibly much less, than 0.1% of the amount of seed protein mRNA present in the embryos. Since this test used nuclear RNA as its source, the absence of hybridization in the leaf shows that the controlling factor—the reason that leaves do not have seed proteins—lies not only in the amount of mRNA, but also in the amount of mRNA precursors (nRNA). This test does not explain how the nRNA becomes so different in the two tissues. Differen-

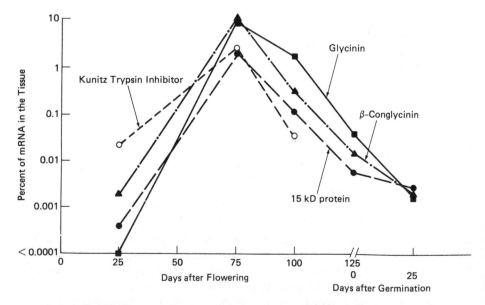

**Figure 3-21.** Accumulation and loss of seed protein mRNAs in soybean cotyledons. Redrawn by permission from R.B. Goldberg, G. Hoschek, G.S. Ditta, and R.W. Breidenbach, *Developmental Biology* 83:218, 1981.

tial synthesis (transcriptional control) is one possibility, but so is differential degradation (turnover) of the nRNAs that are synthesized.

A third experiment directly tested transcription rates by isolating nuclei and measuring the incorporation of labeled nucleotides into cDNA-hybridizable seed protein RNA *in vitro*. The experiment assumes that the rate of incorporation reflects the number of RNA polymerase molecules bound to the DNA and functioning at the time that the nuclei were isolated. A general increase in the amount of labeled, hybridizable RNA occurred during the period of early maturation. There was a decrease between the early and midmaturation phases (Fig. 3–22). These results were consistent with a scheme in which seed protein genes were transcribed in the early maturation phase of embryogenesis; the nuclear RNA was converted to mRNA just before or during the midmaturation phase; proteins were synthesized during the midmaturation phase; and finally the mRNA and the remaining nRNA, if any, were broken down.

Goldberg and Walling have tested the rate of transcription using cDNA clones for several types of seed proteins, including glycinin, beta-conglycinin, and the Kunitz trypsin inhibitor. The fact that many genes for each type of protein are induced at approximately the same developmental stage underscores the importance of coordinated transcriptional control in this developmental process. Harada and Goldberg have also

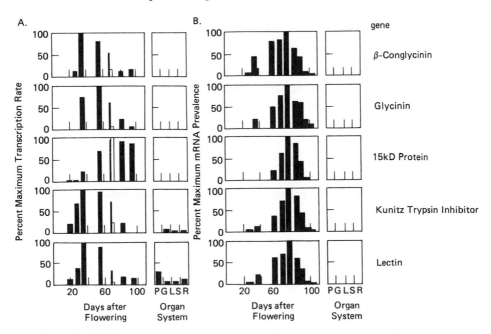

**Figure 3-22.**   Rates of transcription of soybean seed protein nRNAs (left) compared to mRNA concentrations (right) at different stages of embryogenesis. Data kindly provided by R.B. Goldberg.

demonstrated how post-transcriptional modification of nRNA provides a fine tuning for the induction process. Beta-conglycinin genes produce two size classes of mRNA. Transcription of the two size classes occurs at about the same time in the early maturation stage, but one size class (2.5 kb) appears a few days earlier in the midmaturation stage than the other (1.7 kb) size class.

A combination of transcriptional and post-transcriptional regulation has been observed with other legume seed-protein genes. In 1986 Chappell and Chrispeels analyzed the lack of synthesis of phytohemagglutinin in an unusual variety of common bean (*Phaseolus vulgaris*). They found a reduced rate of transcription of the phytohemagglutinin genes (about 20% of that seen in a normal bean variety) and almost no mRNA (less than 2% of normal). Workers at the Commonwealth Scientific and Industrial Research Organization, Canberra, Australia, have been studying the reduction in seed protein content that occurs when pea plants are deprived of sulfur. The reduction is specific for certain high-sulfur proteins, such as legumin. Legumin synthesis recovers if deficient plants are fed sulfate midway through seed development. In 1985, Beach and his coworkers showed that the nuclei of recovering plants doubled their rate of synthesis of legumin nRNA. But the amount of legumin mRNA increases 20-fold. In both of these cases, there is control of transcription and of post-transcriptional processes. It is not known whether post-transcriptional control involves reduced RNA processing or increased turnover. Whatever the mechanism, post-transcriptional control seems to have a strong effect on mRNA supply.

We still do not know how the transcriptional control works, nor do we know what regulates post-transcriptional control. But as we will see in the next chapter, an analysis of the base sequences of different genes, together with their surrounding regions, offers some promise of explaining the control process.

### Technical insert: Determining the Sequence of DNA

Determining the base sequence of a length of DNA was made possible by the discovery of a new experimental strategy that makes DNA pieces into sizes that correspond to the positions of particular bases and then separates the pieces by size. The technique is extremely powerful (it can determine the base sequence of a long piece of DNA in a single experiment) when it is applied to the homogeneous DNA produced by cloning technology.

We will illustrate the operation of one of the two principal methods using a model DNA sequence. This method, originally devised by F. Sanger and his colleagues, makes the DNA pieces by synthesizing them on a homogeneous template.

1. A single-stranded copy of the sequence under study is obtained by cloning it in a variant of bacteriophage M13. M13 is an *E. coli* bacteriophage that has two forms of DNA in its life cycle. The viral particle contains a single-stranded DNA; it is produced from a circular, double-stranded "replicative form" of DNA that is formed in an infected host cell. The "replicative form" can be isolated; and the DNA under study can be inserted in it using restriction enzymes, reassociation of sticky ends, and DNA ligase, just as is done with other cloning vectors. A bacterium that is infected with this hybrid DNA will secrete phage into the medium. The single-stranded DNA that is extracted from this phage contains the DNA sequence of interest (or its complement).

2. The single-stranded DNA is used as a template for DNA synthesis. First, however, it is combined with a short piece of DNA that is complementary to a segment that is adjacent to the insert. This provides a "primer" for DNA synthesis, one with a very specific starting point.

3. The template-primer complex is added to four separate DNA-synthesis reaction mixtures. Each mixture contains DNA polymerase (usually the Klenow fragment, which lacks endonuclease activity), deoxyribonucleoside triphosphates (including one or more labeled with a radioactive isotope), and a certain amount of a 2'-3'-dideoxyribonucleoside triphosphate (ddNTP). The four mixtures contain, separately, ddGTP, ddATP, ddCTP, and ddTTP. The amounts of the ddNTPs are low and are adjusted so that they are incorporated randomly into different places during synthesis. Whenever a ddNTP is incorporated into the DNA, synthesis stops and a fragment of DNA is released. The size of the fragment depends on the point at which the incorporation occurred. Thus, each mixture produces a population of DNAs with a wide range of sizes. These sizes represent the various positions at which the base of its ddNTP was complementary to the template. The four mixtures produce fragments of sizes that correspond to the positions of all four bases.

4. The samples are subjected to electrophoresis in polyacrylamide gel. This separates the pieces by size. If the samples are run in adjacent lanes, it is easy to compare the sizes and thus the relative positions of different bases. The results are visualized by autoradiography.

In Figure 3–23, the numbers refer to the steps described previously.

The second method, devised by Maxam and Gilbert, depends on chemical methods to break DNA into pieces. Assume that the sequence has been isolated by cloning, so that many identical copies exist in solution. The 5' end is labeled with $^{32}$P-phosphate, using T4 polynucleotide kinase and gamma-$^{32}$P-ATP substrate. The DNA is cut with an appropriate restriction enzyme and one of the two labeled pieces is isolated by electrophoresis. This means that only one of the two complementary strands has a radioactive label. Radioactive bands on a gel will refer only to that strand.

The sample is divided into four portions. One portion of the DNA is treated with dimethylsulfate (DMS) and heat. This breaks chains preferentially at G bases. The reaction is adjusted so that there are few breaks per

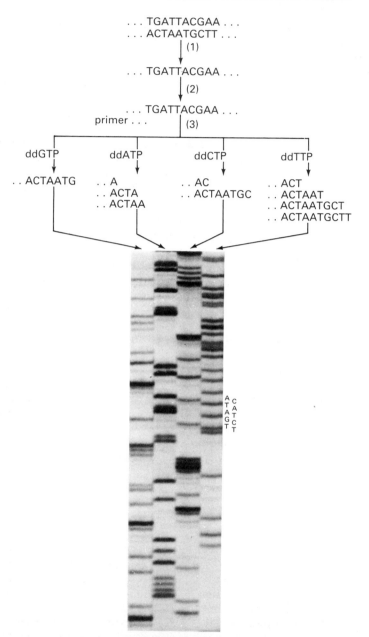

**Figure 3-23.** Determination of a DNA sequence. The diagram at the top shows how, for a short segment of the DNA, the addition of dideoxynucleoside triphosphates can generate polynucleotide chains of different lengths, depending on the base sequence of the template. See text for a description of steps (1), (2), and (3). Photograph by David Stern.

chain. For each G in the sequence, there are some chains that extend from that base to the labeled end. Another portion is treated with DMS plus acid. This breaks chains preferentially at A bases. Yet another portion is treated with hydrazine in the presence of high salt. This breaks chains at C bases. Finally, one portion is treated with hydrazine in the presence of a low concentration of salt. This breaks chains at both C and T bases. The chains broken at T bases are those not broken with hydrazine and high salt.

The four portions are subjected to electrophoresis and autoradiography. The results are read from the autoradiogram, very much as in the Sanger method (Fig. 3–23).

## F. ROOT NODULES AND NITROGEN FIXATION

Most plants obtain their nitrogen from the soil in the form of nitrate and ammonium ions. The nitrogen from these ions cycles through plants, animals, and microorganisms and is released when these organisms die and their constituent chemicals break down. A plant must compete with other plants and with bacteria and fungi for its supply of nitrogen.

In any soil ecosystem, some nitrogen is lost by leaching or by its conversion to $N_2$, which enters the atmosphere. The process of nitrogen fixation, by which atmospheric $N_2$ is converted to ammonia, is one way soil nitrogen is replaced. Nitrogen fixation may occur under natural conditions of high temperature and/or pressure (lightning, thunder, volcanism), but the largest amount by far is catalyzed by the enzyme "nitrogenase," which is found in certain prokaryotic species. Fixed nitrogen that is incorporated into the bodies of free-living nitrogen-fixing prokaryotes is released when these cells die, and a portion of this fixed nitrogen may be used by plants.

Certain species of plants have developed symbiotic relationships with nitrogen-fixing prokaryotes. These plants enjoy a dependable source of fixed nitrogen for which they need not compete. The prokaryotes obtain a source of organic carbon and energy. Some prokaryotes get physical protection as well; they may grow and divide within the plant, being released to infect new plants either gradually or upon the death of the host tissue. The best known symbiosis is that between legumes and species of the bacterium *Rhizobium*. Other nitrogen-fixing symbioses occur between *Alnus* (alder) and *Frankia* (an actinomycete) and between *Azolla* (a water fern) and *Anabena* (a blue-green alga). In addition, at least nineteen other genera of plants have nitrogen-fixing symbioses with *Frankia*.

In the context of this chapter, the legume-*Rhizobium* symbiosis is interesting because it occurs in special organs ("nodules") formed on roots (Fig. 3–24). These organs develop only in response to the presence of bacteria, and they involve the induction of special plant genes. There is like-

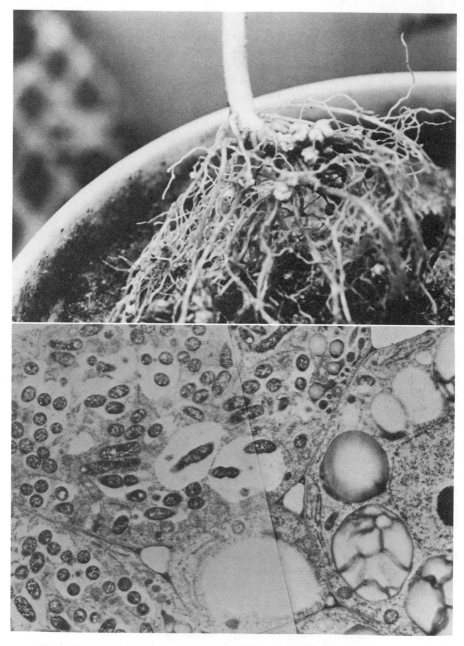

**Figure 3-24.** Top: root nodules on *Phaseolus vulgaris* inoculated with *Rhizobium phaseoli*. Bottom: section of a *Phaseolus* root nodule, showing infected cells (left) with bacteroids inside peribacteroid membranes and uninfected cells (right) with large, conspicuous amyloplasts. Bottom picture reprinted by permission of the University of Chicago Press. From Baird and Webster, Botanical Gazette 143:41-51, 1982. Both photographs courtesy of Dr. Lisa Baird.

wise an induction of bacterial genes, emphasizing the complex, highly evolved nature of the symbiotic relationship.

The formation of a nodule begins as bacterial cells bind to the tip and sides of new root-hair cells. The binding is species specific (that is, clover binds only *R. trifoli,* soybean binds only *R. japonicum,* and so forth). And it is possible to isolate genetic mutants of plants and bacteria that lack binding activity, thus the binding is controlled by genes in both the plants and the bacteria. The binding probably involves an interaction between the polysaccharide coat of the bacterium and a special protein (a "lectin") found on the surface of the root-hair cells.

Once a firm complex has formed between bacterium and hair cell, the hair cell curls near its tip, forming a "shepherd's crook," and the bacterium penetrates the cell wall near the curl. The hair cell forms an infection thread, a tube of cell wall material, which grows through the hair, through the epidermis, and into the cortex of the root. The bacterium moves along this infection thread. As the thread grows, it branches, and the bacterium divides. Eventually many bacteria enter many cortical cells. The bacteria are released from the thread, yet they remain covered with a peribacteroid membrane.

As infection proceeds, the cortical cells of the root divide and expand to form the enlarged nodule. Certain cells may also differentiate to form specialized structures, such as transfer cells (regulating the flow of materials between the nodule and the plant vascular system), aerenchyma (providing a pathway for gas flow), and sclerenchyma (giving the nodule physical protection).

Nitrogen fixation is catalyzed by the enzyme "nitrogenase," and its equation is:

$$N_2 + 12ATP + 6e^- + 6H^+ \rightarrow 2NH_3 + 12ADP + 12 \text{ Pi}$$

This reaction occurs in the bacteria which produce the enzyme. The ammonia is not assimilated into organic compounds in the nodular bacteria, as it would be if they were free-living. Instead, assimilatory enzymes in the bacteria are repressed, and the ammonia is released to the plant cell. There, ammonia is combined with glutamate to form glutamine by the action of the enzyme glutamine synthase. Enzymes convert glutamine nitrogen to other forms. Amino acids are returned to the bacteria for their nutrition, and nitrogen-containing compounds, such as asparagine or allantoic acid (depending on the plant species), are exported to the xylem for transport to the shoot.

Nitrogenase is highly sensitive to oxygen; therefore, an important requirement of the symbiosis is to limit the oxygen concentration. Legume root nodules do this by synthesizing a protein, leghemoglobin, which binds very tightly to oxygen and thus buffers oxygen concentration at a low level. Like animal hemoglobin and myoglobin, leghemoglobin has a heme moiety embedded in a polypeptide chain. The heme is provided by the bacterium; the polypeptide is synthesized from a nuclear plant gene.

From the preceding description, it is clear that the expression of many genes may be regulated during the development of the nodules and the process of nitrogen fixation. In the bacterium, the genes for nitrogenase are induced, whereas those for the enzymes of ammonium metabolism are repressed. In the plant cell, there is a coordinated, specific appearance of leghemoglobin and of the enzymes of ammonium assimilation. It is also possible to identify nodule-specific proteins ("nodulins") as bands in PAGE that are present in nodules but not in uninfected roots. It is possible to identify the leghemoglobin polypeptide (a band of 16kD molecular weight) and, in soybean, a polypeptide subunit of uricase (a band of 34kD). However, in most cases the functional activity of the PAGE-identified nodulins is still a subject for investigation.

Many of the bacterial genes involved in nodule formation and nitrogen fixation are present on a large plasmid, a special circular DNA element that is separate from the main chromosome. This can be demonstrated by growing the *Rhizobia* at high temperature. This causes them to lose the plasmid and also the ability to form nodules. These "uninfectious" *Rhizobia* regain their ability to form nodules when they regain a plasmid through contact with plasmid-containing bacteria. The plasmid has been named "Sym" (for symbiosis).

The Sym plasmid contains two main regions of genes that are essential to development of a nitrogen-fixing nodule. The first region contains two or three genes that control initial steps of infection, such as hair curling, infection thread initiation, and infection thread branching. The regulation of these genes is not understood, because the gene products have not yet been identified. A second region, 45 kilobase pairs away from the first, contains the structural genes of nitrogenase (called *nif*H, D, and K). These genes probably form an operon, the expression of which is controlled like other bacterial operons. In *Klebsiella,* a free-living nitrogen-fixing bacterium, several genes (for example, *nif*A) make positive regulators, which stimulate transcription of the *nif* operon. Several other genes make repressors. For example, the product of *nif*L is a repressor in oxygen and/or fixed nitrogen. Regulatory genes similar to those in *Klebsiella* have been found on *Rhizobium* Sym plasmids by DNA-DNA hybridization.

The induction of plant genes whose expression is nodule-specific has been demonstrated by the work of D.P.S. Verma and his associates. This group isolated poly(A)+ RNA from soybean nodules, synthesized cDNA, and made 5,700 cDNA clones in pBR322. Of these 5,700 clones, 2,100 (37%) were nodule specific, reacting with bulk cDNA from nodules but not from uninfected roots. By looking for hybridization between pairs of clones, Verma identified five major clone types, corresponding to five classes of genes and proteins that are, apparently, those most strongly induced in the nodules. There were 860 clones that contained genes for the 16kD subunit of leghemoglobin. There were 350 clones that contained genes for another gene class, NodA; 55 clones containing NodB genes; 61 clones containing NodC genes; and 6 clones containing NodD genes. The relative numbers of clones reflected roughly the relative numbers of transcripts of each gene class. In all cases the transcripts accumulated only during the development of nodules.

Much more information about the proteins can be obtained by determining the nucleotide sequences of the cDNA clones and making inferences about the proteins' amino acid sequences. Also, by using the nodulin cDNAs to select mRNAs isolated from nodules and then translating the mRNAs *in vitro,* it has been possible to determine the molecular weights of the proteins coded by the mRNAs. These molecular weights may be compared to weights determined for *in vivo* synthesized gene products, thus giving an idea of the post-translational processing steps involved in the expression of the genes.

One example is the analysis of the soybean nodulin-24 (NodC) gene by Katinakis and Verma in 1985. The sequence analysis of the cDNA showed a coding capacity of 15.1 kD (147 amino acids), with the N-terminal amino acids being strongly hydrophobic. The hydrophobic N-terminus suggested a signal sequence which could move the protein across or into a membrane. Confirmation came from the *in vitro* translation of the mRNA selected by hydridization to the cDNA. When the translation was carried out in the presence of dog pancreas microsomes, the product was smaller than it was in the absence of the microsomes. This is a functional test for a signal sequence, since microsomes contain the machinery for binding to a signal sequence, for directing the polypeptide chain into their lumens as it is synthesized, and for then cleaving off the signal sequence. An antibody raised to the peribacteroid membrane reacted with the translation product, indicating that, in the plant, the signal sequence directed the polypeptide chain to that membrane.

An analysis of the cDNA for nodulin-23 (one of the NodA class) also shows a very hydrophobic N-terminal sequence. We can predict that this protein will also be located in a membrane or inside a membraneous organelle, but confirmatory evidence is, of course, necessary.

Not all proteins found inside membraneous organelles have signal sequences. The soybean nodulin-35 (uricase) protein is an example. Antibodies to this protein have been used to show that it is located in peroxisomes of uninfected cells of nodules (Fig. 3–25). However, an analysis of its cDNA indicates no obvious hydrophobic N-terminal signal sequence. The protein is synthesized by free, rather than membrane-bound polysomes, and the size of the *in vitro* synthesized protein is not affected by microsomes. The protein must be taken into peroxisomes after it is

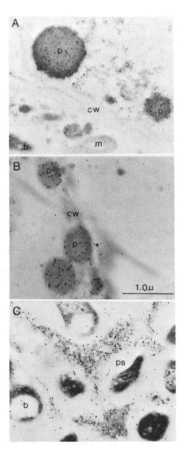

**Figure 3-25.** Immunocytochemical localization of nodulin-35 (a and b) and leghemoglobin (c) in soybean root nodule cells. Dark spots are gold atoms attached to specific antibodies. Legend: p, peroxisomes; cw, cell wall; m, mitochondria; b, bacteroid; ps, peribacteroid space. Reprinted from Nguyen and others, *Proc. Natl. Acad. Sci.* USA 82:5040, 1981. Photograph kindly supplied by Dr. D.P.S. Verma.

synthesized. Other examples of proteins that move into organelles will be discussed in the next section.

All the available evidence suggests that nodule-specific proteins appear through regulation of the amount of mRNA, possibly through regulation of the transcription rate. It is interesting to speculate on the mechanism by which transcription of different genes could be turned on at the same time. In this regard, it is significant that three genes, one for leghemoglobin, one for nodulin-23, and one for nodulin-24, have in common in their promoter regions three sequences totalling more than 40 base pairs. In 1985 Mauro and his associates, who reported this discovery, calculated that the probability of this correspondence occurring randomly is less than $4 \times 10^{-11}$. More extensive discussions of promoters and other regulatory sequences are given in Chapters 4 and 6.

## G. CHLOROPLAST PROTEINS

The most important biochemical system of higher plants—the photosynthetic pathway—is located in the chloroplast. This system is the key to the input of carbon and energy into the plant; and, from an ecological standpoint, photosynthesis supplies carbon and energy to almost all other organisms, too. There are over a score of enzymes and other proteins on which photosynthesis directly depends. One protein in the system, ribulosebisphosphate carboxylase-oxygenase (rubisco) is present in very high amounts in photosynthetic organs (up to 50% of the total protein), and, therefore, it plays a major role in the nutrition of herbivores.

From the standpoint of plant development, there are many reasons to study chloroplast proteins. First, their appearance is tissue-specific. With some exceptions, the enzymes of photosynthesis are found only in leaf mesophyll and stem cortical tissues—green tissues. Second, their appearance is further controlled by environmental factors. Some proteins in some plants depend on illumination. This effect is mediated by phytochrome, at least in part, so a study of the light control in the synthesis of these proteins may help us understand the mechanism of this important growth regulator. Third, chloroplast proteins exist within a membrane-bound compartment of the cell. Some of them are synthesized outside this compartment in the endoplasm. It is interesting to find out how they move rapidly, specifically, and almost completely into the chloroplast. Fourth, some chloroplast proteins are synthesized inside the chloroplast, using a synthetic system that seems more like that of prokaryotes than that of the nucleus and endoplasm. Studying the synthesis of these proteins allows us to see this system in further detail. Fifth, the synthesis of chloroplast proteins in the endoplasm and in the chloroplast appears to be closely coordinated. The coordination mechanism of two separate synthetic systems that depend on mRNA from two independent genomes may be quite different from the coordination of different genes all of which lie in the nucleus. Thus, it is clear why many biochemists and plant physiologists are interested in the study of chloroplast proteins.

The proteins of photosynthesis are fairly well defined (Fig. 3–26). For convenience, we can group them according to their general functions as follows:

**(a) Antenna proteins.**    These bind chlorophyll a and b as chromophores and absorb light. Because they are closely packed and closely associated with the reaction centers of photosynthesis, they can transfer light energy to chlorophyll in the reaction centers by efficient processes that do not require radiation and reabsorption of photons. There are at

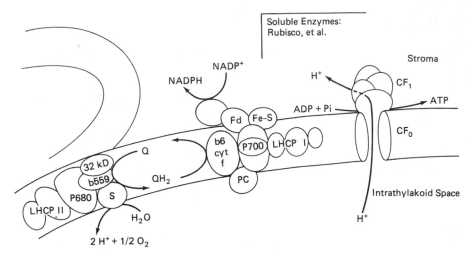

**Figure 3-26.** The protein components of photosynthesis.

least two types of antenna proteins: one for photosystem I and one for photosystem II. They are called LHCP I and LHCP II, respectively, for light-harvesting chlorophyll protein I and II. The LHCPs may each be heterogeneous. These are all found in the thylakoid membranes.

**(b) Photosystem I electron transport proteins.** These include two or three proteins that bind P700 (the chlorophyll molecules that are responsible for charge separation). In addition, certain photosystem I proteins donate electrons to oxidized $P700^+$ (plastocyanin, cytochrome $b6/f$ complex). Also included are proteins that accept electrons from excited P700 (ferridoxin reductase, ferridoxin, ferridoxin-NADP reductase). These proteins are bound to thylakoid membranes, though ferridoxin and ferridoxin-NADP reductase are more loosely bound than the others.

**(c) Photosystem II electron transport proteins.** These include two or three proteins that bind P680 (proteins that oxidize water and reduce oxidized $P680^+$) and proteins that accept electrons from excited P680 for transfer to photosystem I through plastoquinone. The last group includes a 32kD protein that can be inactivated by any of several well-known herbicides (for instance, atrazine and diuron). Cytochrome $b559$ is included in photosystem II, although its exact function is unknown. All these proteins are bound to the thylakoid membrane.

**(d) ATP synthetase (coupling factor).** This enzyme converts the proton and electromotive potential gradient that is produced by the photosynthetic electron transport systems into the free energy of hydrolysis of ATP. This enzyme has two parts. The first part, CF0, is formed from three types of proteins and is buried in the thylakoid membrane, where it forms a pore for the transport of protons. The other part, CF1, is composed of five different kinds of polypeptide chains and is attached to CF0. Thus, CF1 is also attached to the membrane, but it can be removed by relatively gentle treatments.

**(e) Enzymes of the reductive pentose phosphate cycle.** These incorporate and reduce $CO_2$ to carbohydrates, using the energy and reducing power of ATP and NADPH. The enzymes include ribulosebisphosphate carboxylase-oxygenase (rubisco), mentioned previously. This enzyme forms phosphoglyceric acid from ribulosebisphosphate, $CO_2$, and $H_2O$. Other enzymes of this cycle include phosphoglyceric acid kinase, glyceraldehydephosphate dehydrogenase, triose phosphate isomerase, aldolase, transketolase, fructosebisphosphatase, sedoheptulosebisphosphatase, ribosephosphate isomerase, ribosephosphate epimerase, and ribulosephosphate kinase. The enzymes of starch synthesis, including UDPG pyrophosphorylase and starch synthetase, might also be added to this group. All these enzymes are present in the stroma, the solution that surrounds the thylakoids.

As mentioned previously, some of the chloroplast proteins are coded by genes in the nucleus; some are coded by genes in the plastid. This was originally discovered by genetic analysis of plants that contained proteins differing in electrophoretic forms. For instance, electrophoretic variation in the small subunit of rubisco is inherited as a simple Mendelian codominant allele, and it is therefore coded by a nuclear gene. The large subunit of rubisco is inherited maternally, as expected for a gene in the plastid genome. Several other proteins are nuclear, including LHCP, gamma and delta subunits of CF1, CF0 subunit II, plastocyanin, the Rieske protein, and photosystem I reaction center subunit 2. Several other proteins are coded by genes in the plastid, including the 32kD herbicide-sensitive protein, the alpha, the beta, and the epsilon subunits of CF1, three subunits of CF0, cytochrome *b6*, cytochrome *f*, the P700 apoprotein, and two polypeptides of the reaction center of photosystem II.

Although it cannot be taken for granted, it appears to be generally true that nuclear-coded genes are translated in the endoplasm and plastid-coded genes are translated within the chloroplasts. The nuclear-coded gene sequences can be found on poly(A)+ mRNA; no plastid-coded gene sequence is found on poly(A)+ mRNA.

Although there are many different proteins in the chloroplast, only a few of the proteins have been studied in detail from the standpoint of developmental control. These have served as reasonable representatives for the rest.

The entrance of endoplasm-synthesized proteins into the chloroplast has been studied using the small subunit of rubisco as a model system. It was first thought that the enzyme might be synthesized by ribosomes attached to the outer membrane of the chloroplast. And it was thought that the protein was then inserted through the membrane as synthesis proceeded. This would be analogous to the way in which seed proteins are inserted into the lumen of the endoplasmic reticulum. Other mechanisms were also considered. The actual process turned out to be similar to the movement of proteins into mitochondria—the specific uptake of a completed protein in its soluble form (Fig. 3–27). A precursor of the small subunit, containing an extra "transit" sequence of amino acids on its N-terminal end, is synthesized. It is taken up into the chloroplast through the inner and outer membranes by some as yet unidentified carrier. As part of the uptake process, the transit sequence is removed. The transit sequence is therefore like a signal sequence in promoting movement across a membrane except that it works after, rather than during, polypeptide chain synthesis.

This process can be performed *in vitro*, using small subunit precursors synthesized *in vitro* along with purified, intact chloroplasts—this is how it was discovered by Chua's group in New York and Ellis's laboratory at Warwick, England. These investigators showed that the transit sequence was necessary for transport but that the exact amino acids in the sequence were not critical except at the point where the transit sequence joined the rest of the subunit. There is a great deal of difference among the amino acid sequences of transit polypeptides in different species of plants. The exact characteristics that allow the chloroplast to recognize the small subunit from among all the other proteins synthesized in the endoplasm are still unknown.

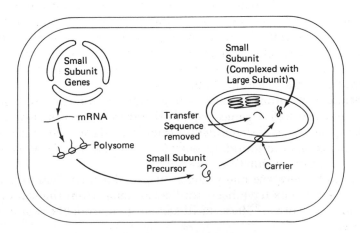

**Figure 3-27.** Synthesis and transport of the RuBP-carboxylase (rubisco) small subunit into the chloroplast.

One of the issues concerning the development of the chloroplast photosynthetic system has to do with its induction by light, the process of de-etiolation or greening. De-etiolation in general involves many changes in a plant, including changes in the shapes of leaves and stems, but the development of the chloroplasts and their photosynthetic ability are among the most striking at the cell level. Physiological experiments suggest that greening may require more than one induction process. For example, small doses of red light do not by themselves make a plant seedling turn green. However, these same small doses of red light do sensitize the plant to synthesize chlorophyll and acquire photophosphorylation ability faster when the plant is placed in continuous white light. This suggests that the red light induces some, but not all, of the processes leading to photosynthetic ability. Red light pulses could stimulate the formation of the $P_{fr}$ form of phytochrome. White light is known to stimulate the conversion of protochlorophyll to chlorophyll.

Several investigators have studied the effect of light on the expression of certain chloroplast genes. Once more the small subunit of rubisco has served as a model. Tobin and her colleagues at UCLA have worked with *Lemna*, a small water plant ("duckweed") that can grow in the dark with a sucrose carbon source, producing white leaves. In the light the leaves turn green. Tobin has extracted poly(A)+ RNA from light- and dark-grown plants and tested for the presence of messenger sequences for the rubisco subunit. The results were the following: (a) Poly(A)+ RNA from polysomes of dark-grown plants translated *in vitro* produces no small subunit. This shows that no small subunit is synthesized because no mRNA is active. (b) Poly(A)+ RNA extracted from total RNA of dark-grown plants, when translated *in vitro,* also produces no small subunit. This shows that no potential mRNA is present in these plants, on or off the ribosomes. (c) Poly(A)+ RNA, as well as total RNA extracted from dark-grown plants, will not hybridize to cDNA clones specific for the mRNA of small subunit. This confirms that the base sequences of the small subunit mRNA are absent in the dark, rather than present but untranslatable. Apparently the level of control involves the appearance or accumulation of mRNA, resulting either from an increased rate of transcription or a decreased rate of RNA turnover. Nuclei isolated from dark-grown plants synthesize extremely low levels of the RNA that hybridizes to small subunit cDNA, yet the RNA that is produced does not break down especially fast. Thus control in the dark seems to result from a lack of transcription.

The characteristics of the small subunit gene that make its transcriptional activity sensitive to light have been studied by altering various parts of the gene *in vitro*. The altered genes are then introduced into new cells where the influence of light on their transcriptional activities can be measured. Chapter 5 describes the results of some of these experiments.

CARL A. RUDISILL LIBRARY
LENOIR RHYNE COLLEGE

The participation of phytochrome was confirmed by showing that a low dose of red light (one minute) was sufficient to increase the amount of small subunit mRNA, and that far-red light reversed the effect of red light. In peas there is also a phytochrome control of the mRNA for the large subunit of rubisco. However, Link has found that in mustard there is no light control of mRNA for the large subunit of rubisco, although light does control the appearance of completed enzyme in this plant. Assuming that it is fair to compare results obtained with different plants in different stages of development, the data suggest that phytochrome can control the appearance of one protein (small subunit) at the level of transcription and another (the large subunit in mustard) at a level other than transcription. The data also suggest that different mechanisms may operate in different plants (peas and mustard) to produce similar end results.

To understand further how light controls the appearance of the large subunit of rubisco, it is useful to consider whether the synthesis of the small and large subunits is coordinated, and, if so, how this takes place. Most experiments to test for coordinated synthesis have used inhibitors, such as cycloheximide, an inhibitor of endoplasmic ribosomes, and chloramphenicol or nigericin, inhibitors of plastid ribosomes. In most cases, cycloheximide preferentially inhibits the synthesis of the small subunit, and chloramphenicol inhibits the synthesis of the large subunit. This by itself suggests that the coordination is not absolute. Indeed, in soybean cells treated with cycloheximide, large subunit synthesis progresses for several hours without any apparent synthesis of small subunit. However, in barley treated with cycloheximide, the apparent synthesis of large subunit falls off with a half-life of two hours. And in *Chlamydomonas reinhardtii* treated with chloramphenicol, there is no apparent synthesis of the small subunit.

The *Chlamydomonas* study is revealing since it was possible to show that even when no small subunit was present the mRNA for small subunit was present and functional. This leads to the suggestion that the apparent coordination of large and small subunits occurs—when it occurs—because those subunits synthesized in the absence of the other subunits are rapidly degraded. This would be an example of control below the level of translation, at the level of protein turnover.

The LHCP is another protein that has been studied extensively. This is a good complement to rubisco, because it is membrane-bound, while rubisco is found in the stroma. LHCP mRNA appears in cells upon illumination and shows the same red–far-red control characteristics as does the small subunit of rubisco. The RNA that hybridizes to LHCP cDNA is synthesized three times faster in nuclei isolated from red-light-treated barley shoots than in nuclei from dark-grown shoots or nuclei from red–far-red-treated shoots. LHCP is synthesized in the endoplasm and is taken up into chloroplasts as a soluble protein precursor, and a transit polypeptide is removed during the uptake. It is not yet known whether the sequence of the transit polypeptide contains some information that guides the insertion of the protein into the thylakoid membrane. Nor is it known whether this function is controlled by the structure of the LHCP after the transit polypeptide is removed or if some other information source is necessary. Red light by itself, while sufficient to stimulate the appearance of mRNA for the LHCP precursor, does not seem to be sufficient to get accumulation of the protein. Some white light is necessary. The white light (or possibly chlorophyll, the synthesis of which is promoted by white light) must in some way control the translation, maturation, or stabilization of the LHCP. By drawing an analogy to the subunits of rubisco, we might suppose that, without chlorophyll, LHCP is rapidly degraded, but this remains to be tested.

It is already clear that the phytochrome control of greening is more complicated than is apparent from studies of rubisco and LHCP. A two-dimensional PAGE analysis of proteins synthesized on polysomes from light- and dark-grown pea seedlings has suggested that at least 35 mRNA transcripts are regulated by light—25 increase and 10 decrease. We can identify two specific proteins that decrease in light-treated plants. These are NADPH protochlorophyll oxidoreductase, an enzyme that participates in the photoreduction of protochlorophyll to chlorophyll, and phytochrome itself. In both cases light stimulates the breakdown of preexisting protein, and it lowers the amount of mRNA available for the synthesis of new protein. With regard to transcripts that are stimulated by light, work at the Carnegie Institution laboratory at Stanford has shown that some respond to a normal "low" fluence of red light, while others, including a transcript for an LHCP gene, respond to a "very low" fluence (almost 10,000 times lower). Some respond to blue light as well as red light; some do not. Some transcripts appear rapidly after a light pulse and accumulate for at least twenty-four hours, whereas others appear only after a lag period of about sixteen hours. Still others appear quickly but transiently. For at least one transcript, stimulation by red light remains reversible for several hours; in other cases, reversibility disappears within an hour. It is difficult to explain all these observations by a single model of phytochrome action, suggesting that there are many complications yet to be unraveled. One possibility, for which there is some evidence, is that phytochrome exists in several forms. These different forms may have different modes of action.

## H. SUMMARY

We have seen several situations in which new gene products appear in response to an internal or environmental signal. In one case, in the chloroplast, the appearance of a protein (the large subunit of rubisco) reflects the stabilization of a polypeptide chain. In the other cases, the appearance of proteins occurs because there are more of their mRNAs available to be translated. In three cases, IAA induction, seed protein synthesis, and light-induced synthesis of rubisco and LHCP, the evidence suggests that the appearance of mRNAs involves an increase in the rate of transcription of the appropriate genes. Much more work will be needed to find out whether these findings can be generalized to other genes.

We might learn more about how gene expression can be controlled by studying the base sequence and organization of genes whose expression is variable. This is particularly true where the control involves transcription. The next chapter addresses this possibility.

# FOUR ————————————————————

# Organization of the Nuclear Genome and Its Genes

## A. GENOME ORGANIZATION

Eukaryotic nuclear genomes can be distinguished from organelle and prokaryotic genomes by their size and complexity. A typical higher plant genome, for example, contains about $5 \times 10^9$ base pairs of DNA per haploid set of chromosomes. This is about 30,000 times as much DNA as in a single chloroplast genome and some 10,000 times as much as in a moderately sized plant mitochondrial genome. And it is 1,000 times as much as in the chromosome of a bacterium such as *E. coli*. Such a large amount of DNA must be very efficiently packaged to fit into the nucleus. Our typical plant genome with $5 \times 10^9$ base pairs of DNA would be about three meters long if all the DNA were somehow laid out in a straight line. In fact, however, the DNA is so greatly condensed in chromatin by being wound around nucleosome particles and by higher order coiling (see Chapter 2, Section B) that it fits into a nucleus about 10 micrometers in diameter. To get a sense of the degree of condensation involved, imagine a nucleus 3 inches (75mm) in diameter, or about the size of a baseball, and then imagine, on a similar scale, that the $5 \times 10^9$ base pairs of DNA would extend for almost 14 miles (22 km).

Although all nuclear genomes are large in relation to organelle and bacterial genomes, their actual sizes vary tremendously. There is some tendency for more complex organisms to have more DNA than simpler ones, but a great deal of variation exists even within groups of similar complexity. Higher plants provide an excellent example of this "C-Value

Paradox," the unexplained variation in the amount of DNA per set of chromosomes. In vascular plants, genome size varies from about $10^8$ to $10^{11}$ base pairs, with many of the higher values being found in supposedly primitive groups, such as the ferns. Even among angiosperms the range is well over 100-fold; differences of 5- to 10-fold are often encountered within plant families and, in a few cases, even within plant genera. Such large differences between closely related species are extremely difficult to reconcile with the notion that all DNA sequences have specifically definable coding or regulatory functions. It seems unlikely that a lily would need 100 times as much DNA as a bean plant, or that *Vicia faba* should have more than five times the DNA of *Vicia sativa*. Instead, as discussed subsequently, it seems more likely that the "excess" DNA in the larger genomes is comprised largely of repeated sequences that are not directly involved in coding or gene regulation.

A distinguishing feature of many eukaryotic genomes is the presence of large amounts of repetitive DNA. As in the case of total DNA content, there is wide variation in amount of repetitive DNA. Some fungi, such as yeast and *Aspergillus,* have so little repetitive DNA that it is difficult to detect by DNA reassociation techniques (see technical insert on p. 120). But most eukaryotic nuclear genomes contain between 10% and 80% repetitive DNA. Figure 4-1 shows that the proportion is remarkably constant in most of the higher plants and animals whose genomes have been studied in detail. Values for repetitive sequence content cluster around 20% for animals and for plants with small genomes and around 80% for plants with genomes above about $10^9$ nucleotide pairs of DNA. The significance of this difference between plants and animals is not clear. We will return to a consideration of evolutionary variation after introducing some of the basic types of repeated sequences encountered in nuclear genomes.

**Types of repeated sequences.**     When eukaryotic genomes were first examined with DNA reassociation techniques it was immediately apparent that they contained sequences varying tremendously in copy number. There was always a portion of the DNA that reassociated at the rate expected for single copy sequences, but often there was also a considerable fraction of the total DNA which reassociated much more rapidly than expected.

Some sequences reassociated so rapidly that rate measurements were impossible to make. This material was later shown to contain *inverted repeat* sequences. As their name implies, these are sequences repeated in inverted orientation with respect to each other (Fig. 4–2). When DNA is denatured and then transferred to conditions in which reannealing can occur, the two repeats on a given strand will reassociate with one another to make an intramolecular duplex. Since the reaction occurs between se-

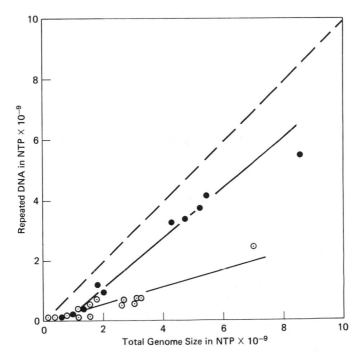

**Figure 4-1.** Amount of repeated sequence DNA as a function of total genome size. Solid circles indicate plant genomes; open circles indicate animal genomes. The dashed line indicates 100% repeated DNA. Data were obtained from studies by many different investigators. See Thompson and Murray, *The Biochemistry of Plants* 6:1, 1981 for a more extensive discussion.

quences on the same strand of the same DNA fragment, it is kinetically zero order (independent of the concentration of the DNA during the reaction) and occurs very rapidly. Experiments have revealed the presence of large numbers of inverted repeats scattered though a large portion of the genome in both plants and animals. Variations in the amount of DNA between the inverted repeat sequences produce a variety of different "hairpin" and "stem-loop" structures, which can be seen when annealed DNA is viewed in the electron microscope.

A great deal of the repeated DNA in higher plant genomes exists in the form of "tandem arrays" in which a basic sequence is repeated serially many times to form a large block. The first examples of tandemly repeated sequences to be studied were the so-called satellite DNAs, which formed small bands separated from the rest of the DNA on isopycnic density gradients. Since very short sequences are more likely to have an unusual base composition and a correspondingly unusual density, the repeating units in satellite DNAs tend to be quite short. For example, a prominant satellite DNA in wheat and barley is composed of long tandem

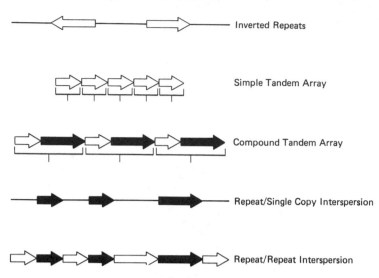

**Figure 4-2.** Schematic representation of different types of repeated sequence organization. Repeated sequences are indicated by the thick arrows. Single copy DNA is indicated by the thin lines. Arrows of the same length and shading are homologous; that is, members of the same repeated sequence family. The direction of the arrows indicates the orientation of homologous repeats with respect to one another.

arrays of the sequence GAA, with some regions of GAAGAG. In other cases, however, the repeating units may be much longer. As the length of the repeating units increases, the likelihood of an unusual base composition and density decreases. Thus many tandem arrays do not form satellite bands on density gradients.

The number of sequence copies in these arrays can be very high. Up to $10^6$ copies of a simple oligonucleotide sequence in a satellite DNA is not uncommon. More complex repeating units several hundred nucleotides long may be repeated 10,000 to 100,000 times. As illustrated in Figure 4–2, arrays may be simple—composed of a single type of repeating unit— or compound. In compound arrays each primary repeat consists of two or more different sequences. The compound nature of many plant tandem arrays was first recognized from the results of interspecific comparisons in which it was shown that species-specific repeats were frequently interspersed with repeats common to two different species. Compound arrays were subsequently characterized by restriction enzyme analysis, using cloned copies of individual repeating units to identify complementary sequences on Southern blots. (See the technical insert on genomic cloning at the end of this section.) An example of such a characterization is shown in Figure 4–3.

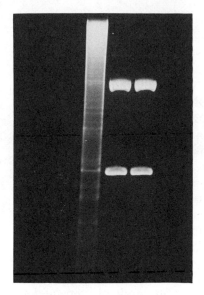

**Figure 4-3.** Structure of a complex tandem array, and a cloned example of a complex repeating unit. The agarose gel on the top shows rye DNA digested to completion with *Eco*R I (lane A) and recombinant plasmids containing a 2.2 kb insert (lower band) corresponding to the major 2.2 kb *Eco*R I repeat in the rye genome. The structure of the 2.2 kb repeat is shown on the bottom. The thick line represents a sequence that was cloned separately and shown to be repeated elsewhere in the rye genome as well as in the 2.2 kb repeat. Reprinted by permission from Bedbrook, and others, and *Nature*, Vol. 288, p. 133, Copyright © 1980 Macmillan Journals Limited.

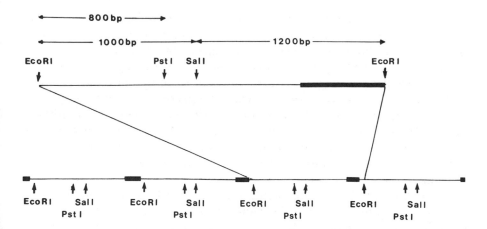

Numerous repeated sequences are "interspersed" in many regions of the genome rather than concentrated in arrays at one or two sites. Some examples are shown schematically in Figure 4–2. It is most common in the literature to consider repeats interspersed with single copy DNA, and, indeed, this is the most frequent configuration for genomes where the repeat sequence content is low. However, in the larger plant genomes where there are higher percentages of repetitive DNA, it is likely that interspersed repeat sequences will be found adjacent to other repeats at least as often as they will be found next to single copy DNA.

**Mechanisms of DNA amplification.** The origin of repeated sequences with features as different as those of tandem arrays and interspersed repeats would seem likely to involve various sorts of sequence amplification events. And, indeed, several different types have been observed experimentally. Tandem arrays, especially those composed of rather simple sequences like the satellite DNAs, may originate through a process of unequal crossover, in which DNA exchange occurs between out-of-register chromatids during mitosis or meiosis (Fig. 4–4). Computer simulations carried out by G.P. Smith showed that such a process could create and maintain tandem repeat arrays. Within an existing array an unequal crossover event will cause one of the resulting sister chromatids to contain more, and the other sister chromatid to contain correspondingly fewer, copies of the repeating unit. Thus unequal crossover can lead to both increases and decreases in copy number. Both increases and decreases have been observed to occur in the evolution of several well-studied tandem arrays, such as those containing the genes for ribosomal RNA, and it may be reliably inferred that unequal crossover is an important source of evolutionary variation in most, if not all, tandemly repeated sequences.

Another way tandem arrays might be generated involves a circular extrachromosomal DNA molecule (such as might be created by intrachromosomal recombination between direct repeats on the same DNA strand) replicating as a "rolling circle." This would lead to the production of long linear molecules made up of head-to-tail repeats of the parental sequence. Such a process is known to occur during the production of the highly amplified extrachromosomal ribosomal gene sequences in amphibian oocytes. If integration of the newly replicated DNA into a chromosome were to occur with sufficient frequency (which need not be very high) this type of amplification could also be a major evolutionary force creating tandem arrays.

Another mechanism involving aberrant DNA replication has been proposed recently by a group working in the laboratory of Richard Axel at Columbia University. They studied the amplification occurring in mouse tissue culture cells transformed with a mixture of plasmids designed to facilitate the subsequent analysis of the amplified DNA. The starting cells were deficient in thymidine kinase, and some of the plasmids contained a thymidine kinase gene modified to reduce but not eliminate its activity. Thus it was possible, by selecting for cells with high levels of thymidine kinase activity, to pick out cells in which an amplification event had produced multiple copies of the thymidine kinase gene. Several plasmids containing cloned human DNA sequences were also included in the transforming mixture, and it was found that some of these sequences were integrated adjacent to the thymidine kinase genes and amplified along with them. (Note that the role of the selection procedure was simply

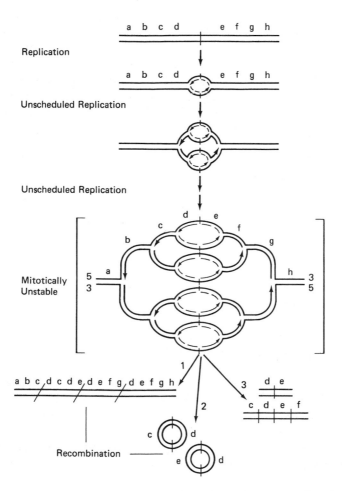

**Figure 4-4.** A model illustrating some of the consequences of unscheduled DNA synthesis and recombination. Normal bi-directional DNA replication generates a bubble, which may undergo additional rounds of abnormal replication to produce mitotically unstable "onion skin" structures. If different replication forks approach each other very closely within such a structure, some linear DNA fragments might be released as extrachromosomal molecules (pathway 3). DNA can also be released in the form of circles by recombination between different homologous sequences within the onion skin (pathway 2). Recombination can also generate amplified arrays of sequences within the chromosome (pathway 1). Reproduced by permission from Stark, *Annual Review of Biochemistry,* Vol. 53, p. 447, © 1984 by Annual Reviews Inc.

to identify those cells having experienced an amplification; the amplification events themselves presumably occurred spontaneously.)

By preparing a series of overlapping genomic clones and using the original plasmid clones as probes, Axel's group characterized one such amplified structure in detail. They compared the copy number and arrangement of plasmid and human sequences in the original transformed cell line with its amplified derivative. Their results showed that a large array consisting of various elements tandemly repeated in irregular order had formed during the amplification. Their data is best interpreted in terms of a model involving repeated rounds of "unscheduled DNA replication" and recombination between homologous sequences within the original transforming DNA. A schematic version of this "onion skin" model is shown in Figure 4–4. An attractive feature of this model is that it can easily explain the widespread occurrence of the compound repeats (Fig. 4–2 and 4–3) that are a prominent feature of highly repeated plant DNA. If the original amplified unit consisted of single copy DNA and two copies of an interspersed repeat sequence, then the product would be a mixture of more highly repeated elements (the original repeats) and repeats of the formerly unique sequence between them. With several repeats in the original sequence the recombinations required to "resolve" the structure could give rise to a bewildering array of different combinations of more and less repeated DNA.

The operation of the mechanisms described previously could give rise to an interspersion of repeat units of differing copy number, but it would not disperse an individual family of repeat sequences over long distances or from one chromosome to another. However, repeat families are frequently dispersed throughout the entire genome in just such a pattern. This has been established most rigorously by the work of Flavell and his colleagues at the Plant Breeding Institute in Cambridge, England. Using *in situ* hybridization of cloned DNA containing several different highly repeated sequence families from the rye genome, the Flavell group showed that copies of repeated sequences were represented on all the rye chromosomes. Similar results were obtained by hybridization to Southern blots of DNA from a series of wheat lines containing individual rye chromosomes (Fig. 4–5). In the latter experiments it was also possible to show that, although family members were present on all chromosomes, some chromosomes contained size classes of repeated sequences not found on other chromosomes. In these cases it is logical to suggest that a particular variant of the repeat, or a combination of a repeat and unrelated DNA, has been amplified in a relatively recent secondary event.

Mechanisms of dispersal may be many and varied. Some may be quite conventional, since it is easy to imagine how such well known chromosomal events as inversions and translocations might result in the distribution of repeat sequences to new locations. However, both the produc-

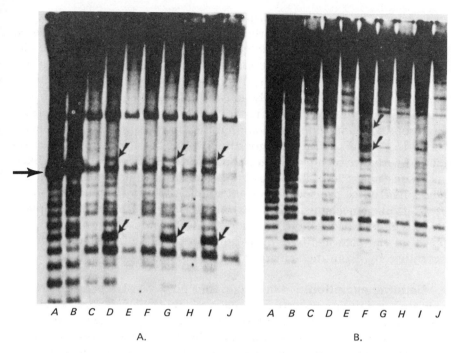

A.                                                    B.

**Figure 4-5.** Southern blots of rye DNA (lanes A and B), wheat DNA (lane J), and seven different wheat lines, each containing a different rye chromosome (lanes C through I). The DNAs were digested with *Eco*R I, fractionated according to size by electrophoresis in an agarose gel, transferred to nitrocellulose, and hybridized with a radioactive probe representing a part of the rye repeated sequence described in Figure 4–3. Because the copy number of this sequence is much lower in wheat DNA and because the wheat background is the same in all the addition lines, most of the bands in the DNA of the addition line plants must represent sequences carried on the rye chromosome present in that addition line. Sequences homologous to the probe are present on all rye chromosomes, and there are large differences in the representation of different variants on different rye chromosomes. Small arrows identify examples of variant bands present on only some rye chromosomes. Reprinted by permission from Bedbrook, and others, and *Nature*, Vol. 288, p. 133, Copyright © 1980 Macmillan Journals Limited.

tion and dispersal of many repeated sequences may also be explained by the activities of "transposable elements." Transposable elements are DNA sequences which, either alone or in combination with other "helper" elements, can replicate themselves independently and insert themselves at new sites in the genome. These elements will be discussed in more detail in Chapter 6. Active transposable elements, or "transposons," have been characterized in several higher plants as well as in bacterial and animal systems. A number of interspersed repeats in plants have been cloned, analyzed in detail, and shown to have structures such as terminal inverted repeats which are found in transposable elements. This observation, together with the presence of a large number of inverted repeats scattered throughout a large portion of the genome, suggests that transposable elements may account for a large portion of the interspersed repetitive DNA. It is important to note that this hypothesis does not require the interspersed repeats to be currently active transposons. It is quite plausible to suppose that much of this fraction of the genome is derived from transposons which have mutated so that they have lost the ability to transpose. Once "fixed" in the genome they would be subject to further sequence divergence by normal mutational processes. Eventually these fixed transposons might lose all recognizable similarity to active elements. Sequences inactive as transposons would still be subject to amplification by a variety of other mechanisms such as rolling circle or "onion skin" replication. It is thus conceivable that a very large fraction of the repeated sequences of the plant genome derive ultimately, though perhaps in many cases indirectly, from the activities of transposable elements.

**Genome evolution.**   The origin and purpose of repetitive DNA is one of the intriguing mysteries of eukaryotic molecular biology. In one of the first well developed hypotheses to explain the presence of large amounts of repetitive DNA in eukaryotes, Britten and Davidson proposed a model in which interspersed repeats formed the basis of a gene regulatory network. Repeats associated with coordinately regulated, but unlinked, genes were assumed to provide sites for *trans*-acting regulatory factors. It is clear that some system with formal properties similar to that proposed by Britten and Davidson must exist. It is also clear, as will be discussed subsequently, that some repeated sequences are important in gene regulation. However, other evidence suggests that the presence and organization of most repetitive DNA cannot be explained in this way. Instead, it seems more likely that those aspects of genome organization are better explained as the product of essentially stochastic events in molecular evolution.

To a large extent this conclusion derives from comparing the DNA content and organization in different but closely related organisms. As a case in point we will consider the pea and mung bean genomes. The mung bean or *Vigna radiata* (familiar to many in the form of etiolated seedlings sold as "bean sprouts") is a member of the legume family, and, as such, it is closely related to peas. Since mung beans and peas exhibit a similar biological complexity there is no reason to suppose that their nuclear genomes should differ greatly in size. However, the pea genome contains about $5 \times 10^9$ base pairs per haploid chromosome set, whereas that of mung bean is about 10-fold smaller at about $5 \times 10^8$ base pairs. Pea is not a polyploid species—indeed, its use in genetic studies since the time of Mendel amply testifies to its diploidy—and a careful analysis of the reassociation kinetics of pea DNA reveals the presence of a fraction which reassociates at exactly the rate expected for sequences present once in $5 \times 10^9$ nucleotide pairs. Thus the difference between pea and mung bean is not simply the result of polyploidy or some uniform replication of all the sequences.

These two genomes also differ in the extent to which repetitive and single copy DNA sequences are interspersed with one another. For experimental reasons it is easiest to look at the average length of purely single copy sequence in order to obtain a measure of the average distance from one repeat element to another. As shown in Figure 4–6, reassociation experiments indicate that there are very few long regions of single copy DNA in pea; most are very short, only about 300 nucleotides in length. A quite different pattern prevails in mung bean, however, where a large fraction of the single copy DNA exists as uninterrupted stretches exceeding 6,000 nucleotides in length.

In both pea and mung bean DNA, as well as in most higher organisms, the repetitive sequence DNA includes sequence "families" (groups of similar sequences) in which the precision of repetition differs. Some families include only identical sequences, whereas others are composed of members whose sequences show considerable divergence, presumably as the result of accumulating different mutations in different family members. The amount of divergence can be determined by reassociation kinetics. Reassociation of the DNA from such divergent families will produce duplexes with mismatched base pairs. Too much mismatch can prevent

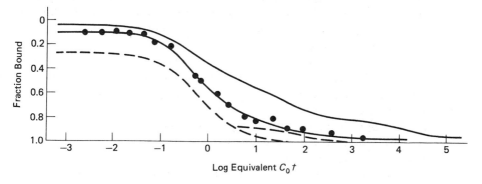

Log Equivalent $C_0 t$

**Figure 4-6.** Reassociation kinetics for long fragments and short fragments of pea DNA. Radioactively labeled DNA fragments with an average single strand length of 1,300 nucleotides were mixed with a 2,000-fold excess of 300-nucleotide long unlabeled fragments, and the reassociation kinetics of the mixture were determined by the hydroxyapatite method (see Fig. 4–9). The solid line depicts the behavior of the short fragments, whereas the solid line with the data points shows the results for the long tracer fragments. Dashed lines represent a computer model of the tracer reaction in terms of two theoretical second order components. From Thompson and Murray, *The Biochemistry of Plants* 6:1, 1981.

reassociation, and the amount of mismatch which can be tolerated is a function of the reassociation conditions. Thus, under relatively stringent reassociation conditions, divergent repeats may fail to form duplexes with each other, whereas their less divergent counterparts reassociate normally. Sequences which fail to reassociate with divergent members of the same family will eventually reassociate with their exact complements, but since there is only one exactly complementary strand per genome these reactions will proceed with single copy kinetics. The fact that the sequences involved really belong to a divergent repetitive sequence family will not be discovered unless the reassociation is carried out under less stringent reaction conditions (for example, at a lower temperature) so that the formation of mismatched duplexes is permitted.

Such an approach was used to provide a clue to the origin of much of the "single copy" DNA in the pea genome. Figure 4–7 shows that much of the DNA, which previously behaved as single copy sequences under normal reassociation conditions, reassociates like repetitive sequences when the stringency of the conditions is reduced by lowering the temperature. Thus many of the sequences which behave as single copy DNA under normal conditions actually are members of repetitive sequence families. Although too divergent to reassociate under normal conditions, sequences in these families remain similar enough to one another to reassociate at the lower temperature. The results with the pea plant support the idea that most of the "single-copy" DNA fraction originates by divergence of repetitive DNA families during the course of evolution. One would not expect most gene sequences to have evolved in the same way, and experiments with a "cDNA" probe prepared by reverse transcription of mRNA show that the gene sequences represented in mRNA do not contain divergent repeats to any significant extent.

The "single-copy" DNA in mung bean seems to contain a much lower fraction of divergent repeats than does the corresponding DNA in pea. Although a small fraction of mung bean single-copy sequences do reassociate as repeats under low stringency conditions, the majority show virtually the same behavior under both conditions. This, coupled with a much lower fraction of repetitive DNA in the mung bean genome (about 25%) than in the pea genome (about 80%), is consistent with a model in which it is supposed that sequence amplification—and hence the production of diverged repeat sequences—has proceeded more slowly in the evolution of mung bean than it has in the evolution of the pea.

Figure 4–8 illustrates this model schematically. It is supposed that a sort of "genome turnover" is occurring during the course of evolution. New families of repeats are arising by amplification events, while other families are diverging in sequence by base substitutions, rearrangements, insertions, and deletions. Some repetitive DNA is being lost by deletion, also. In genomes in which turnover has been rapid, a much greater frac-

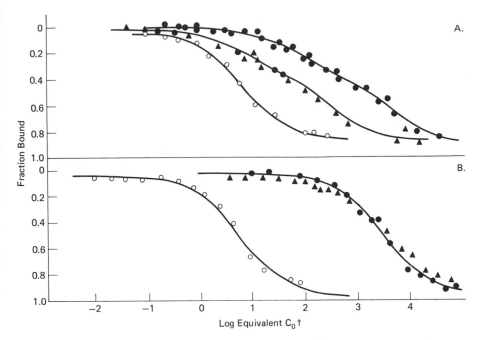

**Figure 4-7.** A. Reassociation of "single copy" DNA and cDNA tracers from pea under normal and low-stringency conditions. The most slowly reassociating fraction of DNA was prepared by preliminary reassociation of 300-nucleotide long fragments to a $C_0t$ value at which most of the DNA had formed duplexes. DNA remaining single stranded at this point was isolated by fractionation on hydroxyapatite. Radioactively labeled fragments from this fraction were mixed with excess unfractionated total DNA and reassociated at either 25°C or 35°C below the $T_m$ of native pea DNA (normal and low-stringency conditions, respectively). The open circles represent the reassociation of differentially labeled *E. coli* DNA included in both reactions as an internal standard. About half of the pea tracer reassociates as would be expected for single copy sequences when the reaction is carried out at normal stringency (filled circles), but essentially all of it reassociates as repetitive sequences when the stringency is lowered (filled triangles). B. As in A. except that the tracer was made by using reverse transcriptase to copy polyadenylated mRNA from pea leaves. In contrast to the pea "single copy" DNA fraction, essentially all of the tracer reassociates as single copy DNA at both normal (filled circles) and low stringency (filled triangles) conditions. From Murray, and others, *Jour. Molecular Evolution* 17:31, 1981.

tion of the total DNA will be composed of repetitive sequences. In these genomes divergent repeats will contribute proportionately more DNA to the "single copy" fraction. Rapid turnover is not the only way to get a large fraction of repeated sequences. A similar result would be expected in a genome having recently been expanded by a series of amplification events which were not balanced by deletions. Either of these scenarios might explain the situation in the pea genome. On the other hand, where

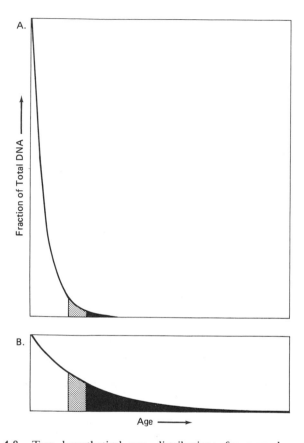

**Figure 4-8.** Two hypothetical age distributions for secondary DNA in genomes characterized by different rates of amplification or turnover (amplification balanced by deletion). Sequence divergence is assumed to accumulate as a linear function of the age (time since amplification) of a repetitive DNA family. The area under the two curves (total genome size) is the same in panels A and B, but the rate of turnover is 4-fold higher in panel A. As a consequence, in case A a larger proportion of DNA consists of relatively young repeats, whereas in case B there is more DNA that is relatively old and has had time to diverge. Sequence divergence will cause DNA from the older repeat families (black) to reassociate as single copy sequences. Younger repeat families (white), having diverged less, will reassociate as repeats. "Fossil repeats" (gray) reassociate as single copy sequences under normal conditions but as repeats when the stringency is lowered. From Murray, and others, *Jour. Molecular Evololution* 17:31, 1981.

amplification has been less frequent and/or occurred less recently, one would predict that a larger fraction of the total DNA would have diverged to the point at which it behaves as single copy sequences even under low stringency conditions.

The properties of the DNA sequences we have been discussing are not the properties one would expect for structural genes. Although introns (discussion follows) can sometimes vary quite rapidly in evolution and may sometimes contain repeated elements, the actual coding sequences generally evolve slowly, are present in only one to a few copies per haploid genome, and form precisely paired duplexes upon reassociation. Very little of the DNA in a typical eukaryotic genome has these properties. For example, less than 2% of the pea nuclear DNA reassociates as single copy sequences under low stringency conditions. However, virtually all of the sequences transcribed into mRNA behave as single copy at both normal and low stringency conditions (see Fig. 4–7). One can infer from these data that sequences actually coding for proteins comprise less than 2% of the pea genome.

Two percent of the pea genome is about $10^8$ nucleotides, about equivalent to the entire genome of *Drosophila* or the cruciferous plant *Arabidopsis thaliana*. Therefore, it is not unreasonable to suppose that all the coding and regulatory information required to make a pea plant could be contained in two percent of its genome. The unanswered question is not why the coding fraction is so small but why there is often, but not always, so much DNA beyond that which seems to be required.

In principle, the variation in genome size from organism to organism could simply reflect random fluctuations in the balance between amplifications and deletions. Alternatively, we may suppose that there is selection for different DNA contents in different species. It seems possible that the amount of DNA per nucleus, irrespective of its sequence content, may influence the rate and duration of various developmental processes. There are good correlations between DNA content and features such as cell size, size of tissues and organs, cell cycle time, duration of meiosis, pollen maturation time, and minimum generation time. If these developmental parameters are influenced either directly or indirectly by DNA content, selection for developmental traits will select for particular DNA content as well.

Most likely the situation is intermediate between these two extremes, with some selection for DNA amount superimposed on a good deal of random variation. A great many details need to be worked out before we will be able to give a much more specific answer. For now, the major lesson to be learned is that most higher plants contain quantities of DNA greatly in excess of their needs for coding and regulatory functions, and that the evolution and organization of this "excess" DNA may be quite different than for the genes and regulatory elements discussed in the next section.

## Technical insert: DNA Reassociation in Solution

Much of our understanding of genome organization derives from experiments in which the two strands of double-helical DNA are separated and then incubated under conditions permitting complementary sequences to reform base paired structures. This process is called DNA reassociation. Sometimes it is called annealing or hybridization. It can be carried out with DNA in solution, as described in this section, or with some of the DNA immobilized on a filter as described in the technical insert on blot hybridization. (See Chapter 3, Section C.) Both solution and filter hybridization may also be carried out where one of the reacting strands is RNA instead of DNA, but we will consider only the process which uses DNA alone.

The major steps in a solution hybridization are outlined in Figure 4-9A. They can be described briefly as follows:

1. SHEARING. In most cases of solution hybridization, it is desirable to break the DNA into short fragments to maximize the separation of repetitive and single copy sequences and to minimize the formation of large multimolecular aggregates. This is best accomplished by mechanical shearing. Most experiments designed to examine hybridization kinetics use fragments sheared to an average length of approximately 300 nucleotides, although fragments up to several thousand nucleotides long can be used for specific purposes.

2. DENATURATION. This is the process of separating the two strands of the helix. It is usually accomplished by heating, although exposure to alkaline solutions (exceeding pH 11) will also denature DNA quite effectively. Strand separation results in an increase in the optical density of a DNA solution referred to as a "hyperchromic shift." In controlled thermal denaturation experiments one can obtain a record of the amount of hyperchromicity as a function of temperature; this is a thermal denaturation ("melting") profile. Similar profiles can be constructed using hydroxyapatite fractionation or S-1 nuclease assays (discussion follows) to monitor strand separation.

   The exact temperature at which the strands of a given DNA fragment separate is determined by the base composition of the fragment, with GC-rich DNA "melting" at higher temperatures than AT-rich DNA. The midpoint, or $T_m$, of the thermal denaturation profile is a function of the average base composition of the DNA, whereas the width of the profile—the range of temperatures over which melting occurs—provides an indication of the heterogeneity in base composition. A typical plant DNA with an average base composition of about 40% GC would have a $T_m$ of about 85°C in a buffer such as 0.12 M sodium phosphate and would mostly melt between 75°C and 95°C. The $T_m$ is a sensitive function of the cation concentration up to about 0.5 M, with the $T_m$ higher at higher cation concentrations. Cations reduce the repulsion between the negatively charged phosphate groups in the backbones of the two strands. With less repulsion more thermal energy is required to separate the strands.

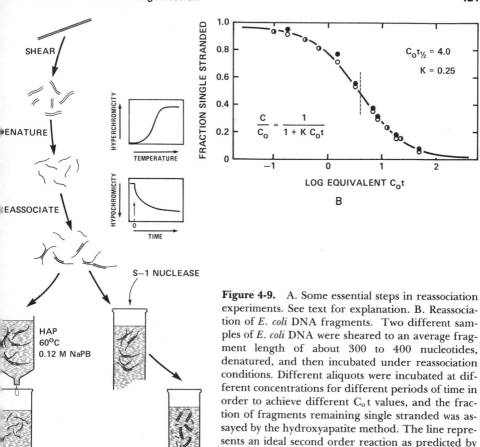

**Figure 4-9.**  A. Some essential steps in reassociation experiments. See text for explanation. B. Reassociation of *E. coli* DNA fragments. Two different samples of *E. coli* DNA were sheared to an average fragment length of about 300 to 400 nucleotides, denatured, and then incubated under reassociation conditions. Different aliquots were incubated at different concentrations for different periods of time in order to achieve different $C_0t$ values, and the fraction of fragments remaining single stranded was assayed by the hydroxyapatite method. The line represents an ideal second order reaction as predicted by the equation.

3. REASSOCIATION. In this process complementary strands anneal with each other to reform duplex structures. If denaturation has been carried out by heating, reassociation can be initiated simply by cooling the sample to an appropriate temperature. For example, DNA in a buffer such as 0.12 M sodium phosphate might be denatured by heating to near 100°C in a boiling water bath and then rapidly cooled to 60°C to start reassociation. A proper choice of temperature and salt concentration is important, as these factors determine the "stringency" of the hybridization conditions. Higher stringency means that more extensive base pairing is required before a particular combination of strands can make a stable duplex. Higher stringency is achieved by raising the temperature and/or lowering the salt concentration. The degree of stringency is most conveniently expressed as "degrees below $T_m$," the number of degrees by which the incubation temperature is lower than the $T_m$ of the native DNA in the particular buffer being used.

The rate of reassociation is optimal about 25°C below the $T_m$. Reactions are faster at higher salt concentrations, but to maintain the same stringency the temperature must also be increased to compensate for the resulting increase in $T_m$. Reassociation of perfectly complementary sequences and of mismatched sequences will have different temperature optima even in the same buffer. Mismatched base pairs lower the $T_m$, and DNA in which mismatch occurs will have a correspondingly lower optimum temperature for reassociation.

Duplex formation in solution is normally assayed by one of three techniques. The simplest is to follow the decrease in optical density which occurs as the result of base stacking interactions in duplex DNA. This "hypochromic shift" is the reverse of the hyperchromic shift used to follow denaturation. The amount of hypochromicity is proportional to the fraction of nucleotides paired at any given point in the reassociation process. Since data can be recorded continuously, this technique permits the most accurate kinetic measurements. However, reassociation of single copy sequences in large genomes requires higher DNA concentrations than those for which optical densities can easily be measured. Furthermore, reassociation of highly repeated sequences requires DNA concentrations that are too low. Measuring optical density is, therefore, inaccurate when the DNA comes from large, complex genomes. Also, optical techniques do not permit the use of radioactive tracers.

Tracers can be used in both hydroxyapatite and nuclease techniques. Hydroxyapatite ("HAP") is a form of calcium phosphate crystal which, under proper conditions such as 0.12 M sodium phosphate buffer at 60°C, will bind duplex DNA but not single strand DNA. Thus, if a mixture of single- and double-stranded molecules is passed through a column of HAP, the double-stranded pieces will stick to the column while single-stranded fragments pass through. Double strands can then be eluted by raising the temperature or the phosphate concentration. In nuclease assays advantage is taken of enzymes such as the S-1 nuclease of *Aspergillus oryzae*, which specifically degrade single-stranded but not double-stranded DNA. The fraction of DNA in duplex structures can thus be determined by measuring the fraction that remains precipitable in trichloroacetic acid after enzyme digestion. Note that the HAP and nuclease techniques measure slightly different things. Single-stranded "tails" attached to double-stranded regions will be retained on a HAP column when the double-stranded DNA binds but will be digested by S-1 nuclease along with fully single-stranded DNA. HAP thus measures the fraction of DNA fragments containing some duplex structure, whereas nuclease digestion measures the fraction of nucleotides actually paired.

When assayed by HAP binding, reassociation of a simple, nonrepetitive DNA exhibits second order kinetics which can be described by the following equation:

$$C/C_o = (1 + KC_o t)^{-1}$$

In this equation $C_o$ is the initial concentration of single-stranded DNA in moles of nucleotides per liter, C is the concentration of free single-stranded

DNA (moles of nucleotides per liter) after t seconds of reassociation, and K is the rate constant. Reassociation kinetic data are usually presented in the form of "$C_0t$" curves ("$C_0t$" being pronounced "Cot" for convenience), in which the fraction of DNA remaining single-stranded is plotted against a log scale of $C_0t$. An example of such a curve for *E. coli* DNA is shown in Figure 4–9B.

Under a given set of conditions the rate of reassociation is proportional to the concentration of complementary sequences. When repetitive sequences are present as in the genomes of higher plants, they reassociate more rapidly than single-copy sequences in the same genome, because they are present at higher concentrations. Given the rate of reassociation of a particular repetitive sequence family relative to the rate for single-copy DNA in the same genome, one can estimate the copy number of the repeats. Most eukaryotic genomes contain many different repeat families with different copy numbers and therefore exhibit complex reassociation kinetics such as those illustrated for pea DNA in Figure 4-6. Such kinetics can generally be modeled by fitting two or three "second order components" representing idealized frequency classes; however, one should always recognize that this is a mathematical abstraction and not an accurate description of the sizes and repeat frequencies of the DNA classes.

### Technical insert: Genomic cloning

Before genes can be studied in detail they must be cloned. This normally means making a clone "bank" or "library" containing enough clones to cover the entire genome several times and then testing them to find the desired gene. This can be a major undertaking if the genome size is very large. The number of clones (N) that must be tested to obtain a probability (P) of finding a particular single copy sequence that represents a fraction (f) of the genome is:

$$N = \ln(1 - P)/\ln(1 - f)$$

To have a 99% chance of finding a given gene $2 \times 10^4$ base pairs in length in the pea genome ($5 \times 10^9$ base pairs) requires the testing of about $10^6$ clones.

When so many clones must be tested, it is clearly desirable to use cloning vectors that (a) give the highest possible number of clones per microgram of input DNA, (b) permit the cloning of large pieces of DNA so that fewer clones will have to be screened, and (c) make the screening process as easy as possible. Genomic cloning work is therefore normally done with one of a number of vectors based on the bacteriophage *lambda*. Specially designed *lambda* vectors will carry fragments up to about 20,000 base pairs in length, and as explained below this system offers much greater cloning efficiency and easier screening than plasmid vector systems.

The life cycle of a *lambda* bacteriophage starts when the phage particle infects an *E. coli,* transferring its DNA into the bacterial cell. Once in the cell the phage DNA may be stably maintained as part of the host genome (in which case the strain is said to carry a *lambda* lysogen), or it may replicate rapidly and be packaged into phage particles. The production of phage particles causes the host cell to lyse; the particles are then released. This process eventually leads to lysis of all the cells in the culture. If, however, the host cells are embedded in agar, the lysis will proceed outward from the original infected cell in such a way that clear plaques become visible against a turbid background of uninfected *E. coli* cells.

Some important elements of a typical *lambda* vector are diagrammed in Figure 4–10. These include the *cos* sites, which are required for DNA to be properly packaged into a phage particle, and a nonessential region of phage DNA (called a "stuffer fragment"), which can be replaced by the DNA to be cloned. The figure shows some of the steps in preparing a genomic library using such a vector. Vector "arms" are prepared by cutting the phage DNA with an appropriate restriction enzyme and separating the large fragments from the smaller stuffer. It is *crucial* that this separation be complete, since otherwise one will obtain a high background of clones containing the stuffer fragment rather than the DNA of interest.

One next prepares the plant DNA to be cloned. Normally this means using an appropriate restriction enzyme to digest, to different extents, several aliquots of DNA. Then the DNA fragments of a suitable size are isolated from each aliquot. This procedure avoids restriction fragments too small or too large to clone efficiently. Of course, the enzyme must be chosen to be compatible with the phage vector being used so that ligation can be carried out efficiently.

Once suitable vector and plant DNAs have been prepared they are mixed together and treated with DNA ligase to produce large multimers of linked phage and plant DNA fragments. Although the ligation occurs at random, sometimes a plant DNA fragment of appropriate size will occur between the "left" and "right" *lambda* arms so that a stretch of DNA resembling a normal phage is produced. Such structures can be "packaged" using extracts from infected *E. coli* cells. Enzymes in the packaging extract cut the large multimers at the *cos* sites and assemble the resulting DNA fragments with phage coat proteins. If the DNA between the *cos* sites is the right size and the arm sequences are intact, infectious particles are produced in which plant DNA fragments replace the stuffer fragment of the original phage. These particles constitute the initial library.

Such a library can be screened directly or subjected to one or more rounds of amplification before screening. Amplification consists of infecting fresh host cells with the library mixture and recovering the phage particles resulting from one round of phage growth. In working with amplified libraries it is wise to exercise caution, since some clones may be lost or altered during the amplification process. At a minimum, clones from such libraries should be "verified" by comparing their restriction maps with those of uncloned DNA on Southern blots made by using several different restriction enzymes.

If the right kind of filter paper is carefully laid over the plaques, some of the phage particles in each plaque will stick to the filter. The filter will

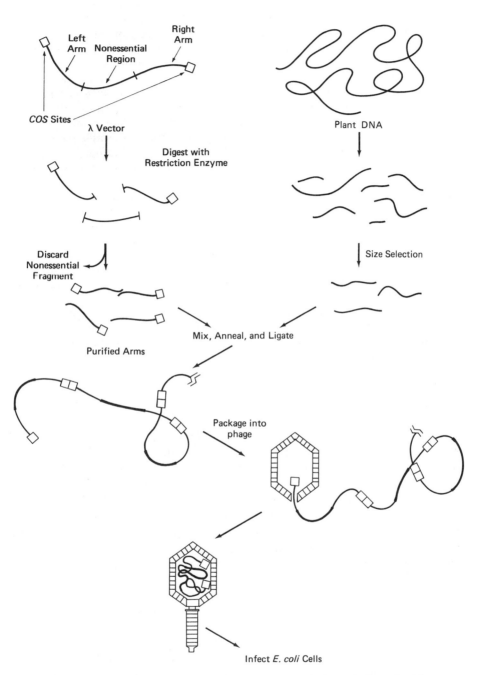

**Figure 4-10.** Steps in preparing a library of clones in bacteriophage *lambda*. See text for explanation.

thus contain a "replica" of the pattern of plaques. After alkali treatment to remove the protein and denature the DNA, it is possible to hybridize such filters with a radioactive DNA or RNA probe much as one would hybridize to a Southern blot (see the technical insert on blot hybridization in Chapter 3, Section C). Autoradiography of the filter then shows the location of any plaques which contain DNA homologous to the probe. One can then go back to the original plate and "pick" viable phage particles for further propagation. In practice, several rounds of phage growth, plating, and hybridization are usually necessary before a pure preparation of a single phage clone is obtained. These steps are referred to as "plaque purification."

If a nucleic acid probe for the gene of interest is not available, it is sometimes possible to use "expression vectors" which have been designed so that the host cells produce a protein whose amino acid sequence is specified, in part, by the DNA in the insert. Protein produced in this way can be adsorbed from plaques onto filters which are then screened with specific antibody. Positive plaques are identified by their ability to react with the antibody, which can be "tagged" in various ways with radioactive or histochemical labels. Thus it is sometimes possible to isolate a gene without going through the intermediate step of making cDNA clones, provided that good antibodies are available for the protein of interest.

Once selected by screening and plaque purification and verified by comparison to genomic DNA, phage containing a gene of interest can be grown to produce large quantities of DNA to serve as a starting point for further analysis.

## B. GENE STRUCTURE

Cloning technology makes it possible to isolate and study individual genes and to ask questions about their structure, regulation, and biological effects. Many nucleotide sequences have now been determined for both plant and animal genes, and our understanding of the relationships between DNA sequences and biological functions has increased rapidly. Most of this work has been done with animal genes, but sequences for plant genes are rapidly appearing, and it is possible to start comparing plant genes to animal genes and to one another. In this section we will review aspects of plant gene structure and function that have been derived from such comparative analysis.

Techniques beyond comparative analysis are required to prove that a given sequence has a particular function. Sometimes it is possible to correlate gene structure and function by analyzing randomly generated mutants, by looking for cases in which an alteration in gene expression cosegregates with a change in the nucleotide sequence of the gene or its adjacent sequences. Although quite successful in a number of microbial systems, including yeast and *Chlamydomonas,* this approach is difficult and time consuming when applied to higher plants with long generation times and large genomes. Alternative approaches involve modifying a cloned gene in the test tube and then testing its ability to be transcribed in an *in vitro* transcription system or its ability to be expressed *in vivo* following transfection (introduction of a modified gene into cells) or following transformation (stable integration of a modified gene into chromosomal DNA). These approaches are being vigorously pursued by a growing number of laboratories, and it seems certain that our understanding of structure and function relationships in plant genes will increase rapidly.

Figure 4–11 is a diagram of a "typical" gene indicating some of its major sequence elements. The following will review what is known about these elements and their actual or potential functions. We will then discuss the way different genes are arranged in the genome.

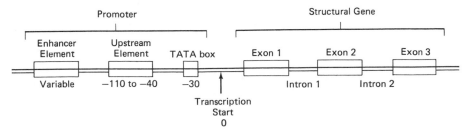

**Figure 4-11.** Diagrammatic representation of a typical nuclear gene. See text for a discussion of the various elements.

**Introns and exons.**    Although there is a wide range of variation, mRNA molecules average 1,000–1,500 nucleotides in length. For a long time it was thought that the average gene must therefore be close to this size as well. Soon after the advent of molecular cloning techniques, however, it was discovered that the distance between the beginning and end of a eukaryotic gene can be as much as ten to thirty times the length of the mRNA it encodes. This is because the coding sequences (exons) are interrupted by intervening sequences (introns). As described in Chapter 2, the gene with its introns is transcribed into long molecules of "hnRNA." During processing the intron sequences are removed by cutting and splicing.

Introns are detected as DNA sequences that lie within a gene yet do not appear in mature mRNA. The best way to detect and characterize introns is by comparing the complete nucleotide sequences of DNA to the mRNA using genomic and cDNA clones. However it is also possible to detect most introns by heteroduplex analysis. Cloned genomic DNA is annealed with RNA and the products visualized in the electron microscope. Since the intron sequences are not contained in the RNA they will form single-stranded loops in the heteroduplex molecule. If these loops are sufficiently large (more than 50 nucleotides in length), they can be seen and their approximate positions in the gene mapped from electron micrographs. An example is shown in Figure 4–12.

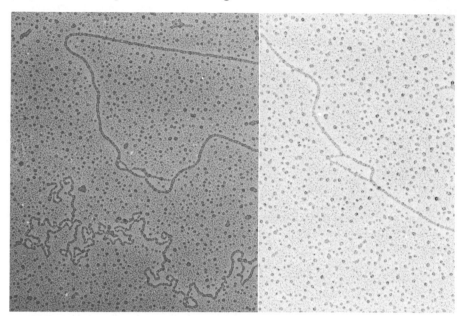

**Figure 4-12.** Electron micrograph of R-loops in soybean beta-conglycinin genes. The gene at the left has introns; the gene at the right has no visible introns. Photograph courtesy of Dr. John Harada.

Although the intron sequences themselves are generally not conserved in evolution, it is striking how often they occur at similar places in homologous genes from different organisms. For example, two of the three introns in the leghemoglobin genes of leguminous plants occur at positions that correspond exactly to the positions of the two introns in the hemoglobin genes of animals (Fig. 4–13A), although the intron sequences themselves have no detectable homology. Another example of conserved intron positions is seen in the alcohol dehydrogenase genes of maize. There are two genes, *Adh1* and *Adh2*, which may have arisen from an ancestral gene duplication event. The exons and the positions of nine different introns are highly conserved, whereas the intron sequences vary considerably (Fig. 4–13B). Essentially the only conserved intron sequences are at the intron/exon junctions. These junction sequences seem to fit fairly closely to the "AG/GT" consensus splice junction sequences identified in animal systems. This probably means that the splicing mechanisms in plant and animal nuclei work in similar ways.

    The evolutionary and functional significance of intron sequences has been much discussed. One of the most appealing hypotheses, originally

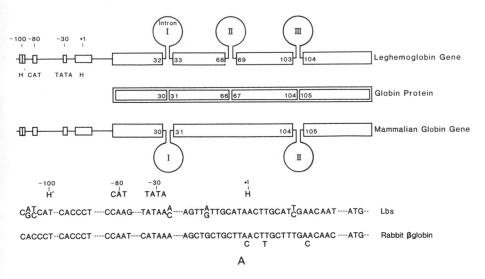

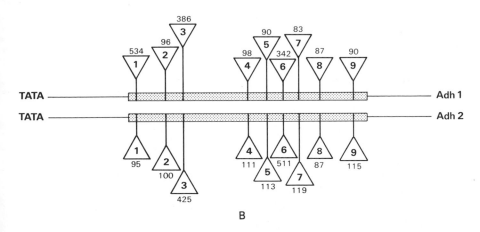

**Figure 4-13.** A. Positions of the introns in leghemoglobin genes of plants compared to those in mammalian globin genes. Shown in the center is a diagram of structural domains of the protein. Introns are numbered with roman numerals. Numbers at the exon/intron junctions are the numbers of the amino acid residues at those positions. B. Positions and lengths of introns in two different *adh* genes. The coding sequences of the genes are indicated by the boxes, which have been aligned at the ATG initiation codon. The positions of the nine introns are indicated by consecutively numbered triangles. Above or below each triangle is noted the length of that particular intron in base pairs. A. is redrawn from Brown, and others, *Jour. Molecular Evolution* 21:19-32, 1984. B. is from Llewellyn, and others, in *Molecular Form and Function of the Plant Genome.* Vloten-Doting, and others, eds. Plenum, 1985.

put forward by W. Gilbert of Harvard University, is that the presence of introns is related to an evolutionary process by which new genes arise from recombination events that juxtapose previously existing genes, or portions of genes. Given the ability to splice transcripts from two or more DNA segments into a single mRNA, "compound" genes might be formed. In support of this idea, exon/intron boundaries in DNA frequently seem to correspond to "domains" or "modules" which are distinguishable in a protein on the basis of secondary or tertiary structure. As one example of this, genes for the small subunit of RuBP carboxylase have an intron at or very near the end of the "transit peptide," which is responsible for directing the protein to the chloroplast. (See Chapter 3, Section G.)

The number of introns is highly variable in both plant and animal genes. There are a number of plant genes with no introns at all, such as the zein storage protein genes of maize and most *cab* genes (nuclear genes encoding the chlorophyll a/b binding protein of photosystem II). Other plant genes have two or three introns. Examples include the leghemoglobin genes mentioned previously, genes for the small subunit of RuBP carboxylase, actin genes, and genes for the legume storage proteins glycinin and phaseolin. A few plant genes have larger numbers and/or longer intron sequences. We have already mentioned the nine introns in the maize *Adh* genes, and the single phytochrome gene so far characterized contains five introns, one of which is about 1,500 base pairs in length.

**Control sequences.**    By comparing DNA sequence data for several different plant genes it is possible to identify a number of conserved regions which, by analogy with animal systems, seem likely to be important for accurate transcription and processing of the RNA. Figure 4–14 shows the results of sequence comparisons by J. Messing and his colleagues at the University of Minnesota. Beginning at the 3' (downstream) end and working backward we first encounter the poly(A) addition signals. Poly(A) is added post-transcriptionally to most eukaryotic mRNAs. In some animal systems it has been shown that a short sequence near the 3' end of the mRNA contains the information necessary for proper 3'-end processing and poly(A) addition. The sequence AATAAA (or AAUAAA in the RNA) occurs in this region in nearly all animal genes examined. By means of *in vitro* mutagenesis experiments, it has been shown to be required for polyadenylation, although other sequences further downstream may also be important. In plants, the sequence comparisons carried out by Messing and his associates identified two sequences with homology to the AATAAA sequence near the 3' ends of several plant genes. Messing and his associates propose that multiple polyadenylation sites occur more frequently in plant genes than in animal genes and that there is more variation in the sequences of these elements in plant genes.

| Clone | Ref | Agga Box | (36-59) | Tata Box | Site ...( ? )...x...( ? )... | Translation Start | Introns..Stop.. | Poly A Additions.............3' Signals |
|---|---|---|---|---|---|---|---|---|
| (g) pSAc3 | 50 | TTTTAAGTGAATCCT | ...(51)... | TCCATACAA TA | ...( ? )...x...( ? )... | AAAAGATGG | ...3.. TAA... | |
| Lba | 30 | | | TCTATATAAACA | ...(32)...x...(49) | GAAATATGG | ...3.. TAA... | (29)AATAAA ...GGATAAA(23/144) |
| Lbc1 | 30 | AGCCAAGAGAAACTT | ...(40)... | TCTATATAAAACA | ...(32)...x...(57) | GAAATATGG | ...3.. TAA... | (29)AATAAA ...GGATAAA(23/144) |
| Lbc2 | 59 | AGCCAAGAGAAACTT | ...(40)... | TCTATATAAACA | ...(33)...x...(54) | GAAATATGG | ...3.. TAG... | (29)AATAAA ...TGATAAA(29/144) |
| Lbc3 | 59 | | | TCTATATAAATA | ...(33)...x...(49) | GAAATATGG | ...3.. TAG... | (29)AATAAA ...GGATAAA( * /131) |
| GmLb11 | 4 | AGCCAAGAGAGACAT | ...(48)... | TCTATATAAATA | ...(33)...x...(49) | GAAATATGG | ...3.. TAG... | (29)AATAAA ...GGATAAA( * /126) |
| Cabb-S | 13 | TCCTAATTGAAATCC | ...(59)... | CCTATTTAAACA | ...(33)...x...(11) | GAAATATGG | ...0.. TGA... | ...CAATAAA(34/295) |
| CMV1841 | 14 | TCCTAATTGAAATCC | ...(58)... | CCTATATAAACA | ...(32)...x...(11) | CAAGCATGG | ...0.. TGA... | ...CAATAAA(34/305) |
| ZG99 | 45 | GCAAAATCGAAAATT | ...(39)... | TGTATAAA TA | ...(30)...x...(57) | CAAGCATGG | ...0.. TGA... | ...AAATAAG(23/87) |
| Z4 | 28 | GCAAAATCGAAAATT | ...(39)... | TGTATAAA TA | ...(29)...x...(57) | CAACAATGG | ...0.. TGA... | ...AAATAAG(23/87) |
| Z7 | 34 | CAAAAGACAAAATC | ...(36)... | TGTATGAA TA | ...(29)...x...(66) | CAATAATGG | ...0.. TAG... | ...AAATAAT(37/97) |
| (c) A20 | 16 | | | | ...x...(68) | CAACAATGG | ...... TAG... | (25)AAATAA ...AAATAAG(23/87) |
| ZG14 | 25 | | | | ...x...(68) | GAACAATGG | ...... TAG... | (25)AATAAA ...AAATAAG(23/87) |
| A30 | 15 | | | | ...... | CAATGG | ...... TAG... | (25)AATAAA ...AAATAAG(23/87) |
| ZG124 | 25 | | | | ...x...(57) | CAACAATGG | ...... TAG... | (25)AATAAA ...AAATAAG(16/80) |
| ZG19 | 25 | | | | | CAACAATGG | ...... TAG... | (25)AATAAA(12-43) |
| B49 | 16 | | | | | | ...... TAG... | AAATAAT(37/97) |
| p222.3 | 38 | | | | | CAACAATGG | ...... TAG... | AAATAAT(17/77) |
| pZML84 | 17 | | | | | | ...... TAG... | (44)AATAAA ...AATGAG(35/372) |
| pRC2.11.7 | 9 | | | | | | ...... TAG... | (30)AATAAG ...AATAAA(23/134) |

| | Agga Box | Tata Box | Start | Poly A Signals |
|---|---|---|---|---|
| Plant Consensus | $^C_T A_{2-6} T\ ^G_T NGA_{2-4} T\ ^{CC}_T$ | $^T_G TATA T A_{1-3}\ ^C_T A$ | $^C_G CAANNATGG$ | $^G_A ATAA_{1-3}$ |
| Animal Consensus | $GG^C_T CAATCT$ | $TATA^T_A^A_A$ | $^A_G NNATGG$ | AATAAA |

**Figure 4-14.** Plant gene structure. Potential control sequences of plant genes are aligned for comparison. Sequences in boldface type agree with the plant "consensus" sequences listed at the bottom of the figure. Numbers in parentheses indicate the distance (in base pairs) between the indicated elements. The first of the two numbers in the parentheses on the far right indicates the distance between the second polyadenylation signal and the 3' end of the mRNA, while the second number is the total length of the noncoding RNA at the 3' end. The first group of clones, labeled "g", are genomic clones; the group labeled "c" are cDNA clones. From Messing, and others, in *Genetic Engineering of Plants: An Agricultural Perspective*, Kosuge, and others, eds. Plenum, 1983.

Actual termination of transcription occurs well downstream of the polyadenylation site in eukaryotic genes (with the possible exception of yeast genes) transcribed by RNA polymerase II. The sequences involved in termination have not yet been well defined even in animal systems, although preliminary evidence from deletion/insertion analysis indicates that well defined terminator regions do exist. One reason why less is known about termination reactions in eukaryotes than in prokaryotes is that most eukaryotic mRNAs are cleaved and polyadenylated at the 3' end. Therefore, the resulting 3' RNA fragment is metabolically unstable and difficult to study. Although this makes termination more difficult to study, the existence of an elaborate 3' processing system may make the actual termination process less crucial, since even transcripts whose 3' ends are not well defined can be cleaved and polyadenylated to produce mRNA molecules with the correct terminal sequences.

Within the coding region plant genes do not seem to differ much from animal genes. Translation-stop and intron-splicing signals are similar, and the initiator codon (ATG in DNA, or AUG in RNA) occurs within a consensus sequence very similar to its counterpart in animal genes.

Most of the transcriptional control sequences are located in the region 5' to the start of transcription. Locating this site is, therefore, an important first step in analyzing sequences in the promoter region. Unfortunately, the position of the translation start codon is not a good guide to the position at which transcription starts since transcribed but untranslated sequences of variable length are commonly found between these two positions.

The procedures illustrated in Figure 4–15 allow one to determine the point in the gene sequence at which the sequence of a particular RNA begins. This point is often, but not always, the same as the start of transcription. Such experiments only locate the 5' end of the RNA as it exists during the experiment. If nucleotides have been removed from the 5' end by processing or by degradation prior to the experiment, the end of the RNA will not be the start of transcription. The actual start of transcription can only be located with certainty if one can demonstrate a "cap structure," formed by the addition of guanine nucleotides to the 5' end of an RNA transcript. (See Chapter 3.) Caps can only be added to RNAs where there is a 5' triphosphate, and 5' triphosphates remain only on the initiating nucleotide in an RNA chain. Since any cleavage or degradation that removes nucleotides from the 5' end of an RNA will also remove the cap and the resulting RNA will not contain 5' triphosphates, the cap structure is diagnostic for the transcription initiation site.

Upstream from the cap site one normally encounters one or two short sequence elements that are common to many eukaryotic genes and that seem to be involved in normal transcription. The best characterized of these is the "TATA box." This sequence, or one very similar to it, oc-

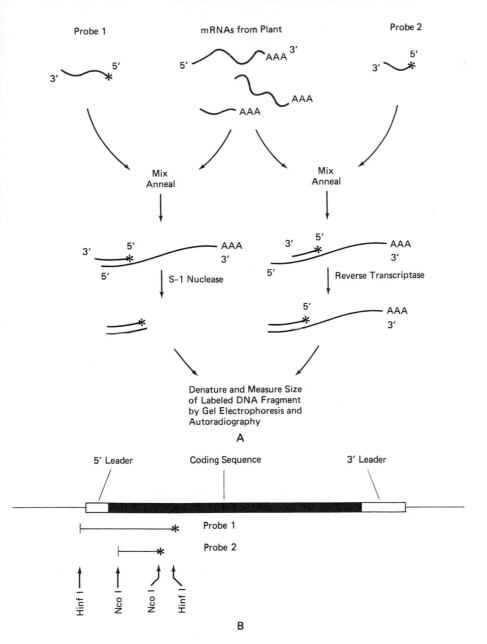

**Figure 4-15.** A. Schematic representation of steps in S1 nuclease and primer extension techniques for mapping the 5′ ends of mRNAs. In both cases probes labeled at the 5′ end with $^{32}$P are prepared so that the labeled end corresponds to a known restriction site in the gene sequence. For S1 nuclease experiments (left) the probe is chosen so that the unlabeled end extends beyond the end of the mRNA, and S1 nuclease is used to digest single stranded regions after annealing the probe with RNA in solution. The length of the protected probe fragment gives the distance between the 5′ end of the probe and the start of the mRNA. Primer extension experiments (right) use a probe that ends within the mRNA sequence. After annealing the probe with RNA, reverse transcrip-

*(continued)*

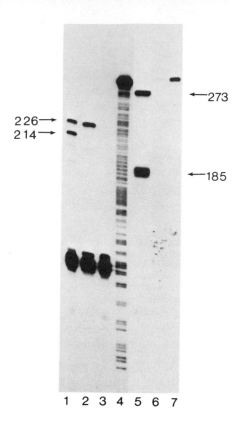

$226 \rightarrow$
$214 \rightarrow$

$\leftarrow 273$

$\leftarrow 185$

1  2  3  4  5  6  7

tase is used to copy the region between the end of the probe and the 5′ end of the mRNA, using the 3′ end of the probe as a primer. The size of the extended primer is then measured as an indicator of the distance to the end of the mRNA. B. Probes used to map the 5′ end of the *cab* mRNA. The positions of the probes in relation to the gene sequence are indicated. Probe 1, used for S1 nuclease mapping, is a *Hin*F I restriction fragment covering nucleotides -20 to +273 in the gene sequence. Probe 2, for primer extension, is a *Nco* I fragment covering nucleotides +96 to +226. C. Analysis of the products of S1 and primer extension mapping experiments with a pea *cab* gene. Lane 4 is a sequence ladder obtained by carrying out the Maxam and Gilbert G+A reaction with Probe 1 and is used to provide markers. Lane 5 contains the products of S1 digestion after annealing Probe 1 with RNA. The band at 273 nucleotides indicates that many mRNA molecules start at a position 20 nucleotides downstream from the unlabeled end of Probe 1 (see B). The band at 185 nucleotides probably results from hybridization of the probe with mRNAs from other *cab* genes whose 5′ leader sequences are not complementary to that of this particular *cab* gene. Lane 6 and 7 contain *E. coli* RNA as a control, and in lane 7 the S1 nuclease was omitted. Lanes 1-3 show the results of primer extensions using Probe 2. Hybridization with RNA was carried out at 45°C, 50°C, or 55°C, respectively. At 50°C, the highest temperature at which hybridization occurred, a single band at 226 nucleotides was detected, indicating that the end of the most homologous mRNA is located 226 nucleotides upstream from the 5′ end of this probe. This position is the same as that identified by S1 analysis. From Cashmore, *Proc. Natl. Acad. Sci.* USA 81:2960, 1984.

curs at about nucleotide -30 (30 nucleotides upstream from the cap site) in almost all plant and animal genes. This element seems to be required for accurate initiation of transcription. A second sequence element, which may be involved in regulating the level of transcription of some genes, is often found a variable distance further upstream, usually between nucleotides -40 and -110. The sequence of this element is much more variable but often includes the sequence CAAT, which molecular biologists have been unable to resist calling the "CAAT box." In plants, the comparisons done by Messing and his associates suggest that this element might be better described as the "AGGA box" since the consensus sequence for their sample of plant genes is somewhat different from that for animals. More information is necessary before the function of this sequence, and its permissible range of variation, are understood.

Other regulatory elements also occur in this region. In general, these elements seem to modify the "promoter specificity," making transcription of the gene responsive to particular environmental signals. Promoter specificity can often be demonstrated by attaching promoter-containing sequences of one gene to the coding sequences of another and introducing the resulting hybrid gene into cells where it can be expressed. Although there are a few cases known where regulatory elements occur in the gene itself (a prime example being 5S rRNA genes transcribed by RNA polymerase III), transcriptional responses usually seem to be determined by the sequences 5' to the start of transcription, often called 5' flanking sequences. A good example is provided by the heat-shock genes which are activated when cells are subjected to thermal stress. (See Chapter 3, Section A.) In both animals (such as *Drosophila*) and plants (such as soybean) it has been possible to show that the 5' flanking sequences from a heat-shock gene can confer responsiveness to a gene not normally responsive to heat shock. In *Drosophila*, the flanking sequences of different heat-shock genes contain a common element that is 14 nucleotides long and located 11–28 bases upstream from the TATA box. This element seems to be required for responsiveness to heat shock and is located in a region which can be shown by direct binding experiments to interact with a protein transcription factor. Similar elements have now been characterized in genes of many organisms, such as yeast and various animals, that are responsive to environmental or hormonal stimuli.

Another most interesting class of control elements is the enhancer sequences. These elements were originally discovered in viral genes, where they were shown to be required for high levels of transcription. They have since been found in a wide variety of animal and some plant genes. Enhancers are typically located within a few hundred nucleotides upstream from the start of transcription, but experiments with genes altered *in vitro* and reintroduced into cells show that enhancer activity is affected very little by changes in position or orientation, as long as the enhancer sequences remain within a few kilobases of the gene. Indeed, the ability to influence gene expression from different positions and orientations is the main diagnostic feature of enhancer elements. Although most enhancer elements are found in flanking sequences on the 5' ends of genes, examples are known in which they occur elsewhere. For example, some mouse immunoglobulin genes have an enhancer embedded in an intron sequence. It is not yet certain exactly what constitutes an enhancer sequence in many systems. However in the prototype enhancer, that of SV40, and in enhancers from several other animal viruses, activity is associated with a set of short tandem repeats 50–100 nucleotides in length. An "enhancer core" sequence such as GGTGTGGAAAG, or more generally GTGGT/AT/AT/AG ("T/A" means that either T or A is found in this position), can be identified by sequence comparisons, although it is clear that sequences other than the core are also important. Enhancer sequences are often associated with regions of alternating purines and pyrimidines which are capable of forming Z-DNA. (See Chapter 2.)

In many cases, enhancer elements seem to be involved in tissue-specific gene expression, acting to facilitate active transcription only in the appropriate cell type. In animal systems it can be shown that this specificity reflects the presence in certain cells of protein factors capable of interacting with the enhancer sequences in question. In plants, enhancer-like sequences stimulate the expression of genes like RuBP carboxylase in light-grown cells but not in cells grown in the dark. (See Chapter 6, Section D.)

**Repeated sequences regulating gene expression.**     In a strict sense, some of the control sequences just described are really repeated sequences. Short elements, such as the TATA or CAAT boxes, appear in many promoters and are therefore repeated, but they are far too short to be recognized in hybridization experiments such as those discussed in the first section of this chapter. Enhancers may sometimes contain several repeats of a somewhat longer sequence (such as the 72-nucleotide repeats of the SV40 enhancer). By themselves, however, these do not reach sufficiently high copy numbers to be detected in classical reassociation experiments, and no extensive sequence homology exists among enhancers as a class, although some short homologies have been found. Therefore, none of the gene regulatory elements we have discussed so far can be considered "classical" repeated sequences.

Evidence for the involvement of "classical" repeated sequences in gene regulation is far more circumstantial and is based on the occurrence of repeat sequence transcripts in the RNA population. Although most mRNAs are transcripts of single-copy genes, repeat transcripts can often be found when cloned repeat sequence probes are used to increase sensitivity. (See Figure 4–7.) This is especially true in nuclear RNAs, where hn-

RNA contains transcripts of many interspersed repeat sequences. When different stages of development are compared, it is usually found that the repeat sequences which are most abundant in RNA differ from stage to stage. Of course, simply because repetitive sequences are transcribed, or because their transcription changes during development, does not prove that they are involved in gene regulation. It seems likely that most transcribed repeats are simply sequence elements that have found their way into transcription units by evolutionary accidents of rearrangement or transposition. On the other hand, it must be admitted that a small class of functionally important repeat sequences might exist against a large background of evolutionary noise. Several lines of evidence from animal systems suggest that this may be true.

One type of repeat of obvious interest is exemplified by the "opa" family of repeats in *Drosophila*. These are sequences homologous to a 93 bp repeat that was identified in clones from the *notch* locus, a gene concerned with neurogenesis. They hybridize with a large family of transcripts that share a common pattern of developmental regulation. It is not known if the "opa" repeats are important in regulating the transcription or translation of these RNAs, but the developmental correlation is intriguing.

One very interesting repeated element is the "homeo box" sequence, which occurs in several "homeotic" genes of *Drosophila*. "Homeo" and "homeotic" refer to the original definition of these genes as loci where mutations caused the replacement of a structure characteristic of one body segment by an homologous, but different, structure from another body segment. An example is the *Antennapedia* mutant, in which the antennae are replaced by a pair of legs from the middle of the body. Homeotic genes show strongly position-dependent patterns of expression, transcripts of different genes being confined to different regions. They are thought to be involved in determining the spatial organization of the developing embryo. The homeo box is a sequence of about 180 bp which occurs in the 3' exon sequences of several of these genes. Its association with homeotic genes is so strong that more than ten different genes with homeotic effects have been isolated from random genomic libraries simply by screening them with homeo box probes. Because of this tight association, it has been thought that the homeo box sequence might somehow be important in identifying groups of genes controlling segmented patterns of development. But it is not yet clear exactly what the function of the homeo box is in molecular terms. Since it is a part of an exon sequence and contains an open reading frame, it may function at the protein level rather than in the control of transcription. It has been suggested that the homeo box sequence might encode a highly basic protein with many of the features of DNA binding proteins. As such, it could function as a "tag" directing the protein to the nucleus or perhaps to specific locations within the nucleus, where it might have effects on the expression of many other genes.

Even more intriguing is the discovery that a group of genes expressed in the rat brain (and specific to neural tissue) contain repeated sequence elements in their introns. These "identifier sequences" can be transcribed by RNA polymerase III to produce two populations of small RNAs about 100 and 160 nucleotides long. Identifier sequences occur elsewhere in the genome, but seem to be transcribed into RNA mainly in neurons. In neural RNA, identifier sequences are present not only in the small RNAs transcribed by RNA polymerase III, but also in large molecules that are transcribed by RNA polymerase II and presumed to be mRNA precursors.

The arrangement and developmental specificity of identifier sequences and their transcripts suggests that they may have a role in regulating neuron-specific genes. J. G. Sutcliffe and his colleagues at the Scripps Clinic and Research Foundation in La Jolla, California have described a model in which the transcription of identifier sequences by RNA polymerase III leads to structural changes in large regions of chromatin, which render the genes within these regions accessible to RNA polymerase II. According to this model, identifier sequences would be involved in a "first round" control mechanism, defining a set of chromosomal regions as potentially active in those tissues or cell lineages that express the proper polymerase III transcription factors for identifier sequence transcription. The result would be analogous to the developmental concept of determination. For example, the range of possible responses to a stimulus would be limited to a particular set that would be defined by the set of genes in the potentially active state. Subsequent regulation of genes within this set would involve RNA polymerase II transcription factors and control sequences such as those discussed in the preceding paragraphs.

The transcriptional elements we have just described are parts of a gene or its flanking sequences, and they affect only the gene to which they are directly linked. In genetic terms, they function *cis*, regulating genes on the same chromosome. Clearly, there must be other elements involved in gene regulation, and in several cases there is now good experimental evidence that certain protein factors must be bound to specific sites in a gene or its flanking sequences in order for transcription to be initiated. An example would be the transcription factor which binds to the heat-shock promoter in *Drosophila*. Genes encoding these factors would be examples of *trans*-acting control elements. As their name implies, such elements are capable of affecting genes other than those located on the same chromosome. *Trans*-acting regulatory genes are well known from genetic studies but our understanding of the molecular nature of such elements, and the factors which presumably mediate their effects, is still quite limited. Certainly, their importance can hardly be denied.

**Ribosomal RNA gene families.**    Probably the best known multigene families are those encoding ribosomal RNA. The "major" ribosomal RNA genes are organized in long tandem arrays containing both gene and spacer sequences. In most animal cells there are some 100–200 rRNA genes, and in plants the numbers can be much higher. It is not uncommon to find 5,000 rRNA genes per genome in plant species with DNA contents typical of crop plants, and much higher numbers have been reported. The genes exist in large tandem arrays at one or a few loci in the genome. The repeating unit in these arrays consists of one large transcription unit containing genes for the 18S, 5.8S, and 25S rRNAs as well as for spacer sequences that are removed during processing of the large primary transcript. Also included in the basic repeat are spacer sequences which are not part of the primary transcript and often called nontranscribed spacer or intergenic spacer sequences. Their importance in the evolution and expression of the rRNA genes will be discussed subsequently. The major rRNA genes are transcribed by RNA polymerase I, a specialized form of RNA polymerase, which transcribes only these genes. Active rRNA genes are found in the nucleolus where their transcripts are processed and assembled with ribosomal proteins.

Genes for 5S ribosomal RNAs are also organized in tandem arrays, although they are located elsewhere in the genome away from the major rRNA gene arrays. They are transcribed by RNA polymerase III rather than polymerase I. As in the case of the major rRNA genes, the arrays consist of alternating gene and spacer sequences. In wheat, two main variants exist in which the repeating units are 410 and 500 bp long. Similar heterogeneity is seen in flax, where the major length classes are 340 and 360 bp. Within these repeats the gene itself occupies only 118 bp. Most spacer sequence is highly variable, even between the different length classes in a single genome, but there is a region of about 70 bp 5' to the start of transcription that is conserved. Studies of 5S genes in *Xenopus*, however, have shown that this region is not required for accurate transcriptional initiation. Interestingly, the promoter sequences required for proper initiation by polymerase III in extracts of *Xenopus* oocytes seem to be located well within the coding sequence of the gene.

The number of copies of both the major rRNA genes and the 5S rRNA genes can vary widely among closely related species of plants, and even among different races or varieties within a species. For example, different lines of flax have been reported to contain between about 50,000 and 120,000 5S RNA genes. The number of major ribosomal RNA genes varied between about 1,400 and 2,700 in the same lines, although there was no correlation between the numbers of 5S genes and major rRNA genes. Extensive variation in the number of major rRNA genes also occurs independently at each of the several loci that contain rRNA genes in hexaploid wheat. Such extensive variation between vigorously growing, closely related genotypes makes it very difficult to argue that the larger gene numbers are required for normal plant growth and development. Instead, it would appear that plants simply tolerate a great deal more variation than animals do. Even in plants with lower gene numbers there may be a substantial "excess" over the number of genes actually expressed in any given cell.

In addition to variation in rRNA gene number, many plants have substantial polymorphism in the length of their tandem repeat units. Both types of variation are thought to result from unequal crossover events. Why both copy number and length variation can result from unequal crossover in ribosomal RNA gene arrays can be understood by considering the structure of a typical repeat in such an array (Fig. 4–16). In each repeat there is the large transcription unit containing genes for the 18, 5.8, and 25S RNAs and the spacer sequences between them. There is also the "nontranscribed spacer." Within the nontranscribed spacer there are a number of short (about 130-180 bp) "subrepeat" sequences. These subrepeats provide another set of tandemly repeated sequences at which unequal crossovers might occur. Unequal crossover in this region would generate spacer length variants with different numbers of subrepeat sequences. This model predicts that most of the rRNA gene length variants seen in nature should differ from each other by lengths corresponding to integral multiples of the subrepeat sequence, and this is exactly what is observed.

The evolution of ribosomal RNA gene families is also characterized by a rapid accumulation of point mutations and other sequence changes in the nontranscribed spacer region. Clearly, the sequence of this region is not conserved in the same way as that in the coding region, which shows strong homology over very large evolutionary distances. In spite of the rapid evolution of spacer sequences, however, a high degree of sequence homogeneity is somehow maintained within the arrays of a particular

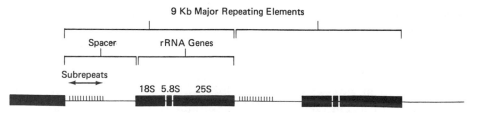

**Figure 4-16.** Map of the repeating units containing the ribosomal RNA genes of wheat. Each repeating unit is approximately 9 kb in length and contains genes for the 18S, 5.8S, and 25S rRNAs as well as spacer DNA. Transcription begins in the spacer DNA just 5' to the 18S gene. In the spacer region upstream (to the left) from the transcription start there is a set of 135 base pair sequences that are tandemly repeated between 10 and 15 times within each of the basic 9 kb repeating units (referred to as "subrepeats"). Among closely related species, most of the variation in rDNA restriction patterns can be attributed to variations in the number of these subrepeat sequences.

genome. This paradox provided the first recognized example of "concerted evolution," in which repeated sequences of multicopy genes sometimes show a tendency to evolve together rather than diverge by accumulating different mutations. Concerted evolution is thought to reflect the operation of homogenizing processes such as gene conversion and unequal crossing over. Gene conversion is the direct conversion of one sequence to another while the sequences are paired during mitosis or meiosis. Unequal crossovers within large tandem arrays, such as those containing the ribosomal RNA genes, can lead to the spread of random sequence variants through the population of genes by a process that is analogous to genetic drift in a population of organisms.

Although it is often thought (and in many cases it is undoubtedly true) that sequences that diverge very rapidly in evolution must be phenotypically neutral or unimportant, the nontranscribed spacer sequences in the major ribosomal RNA genes contain transcriptional promoters and enhancers which are essential to gene function. In contrast to control sequences of protein-coding genes, which can sometimes be recognized by their evolutionary conservatism, promoter and enhancer functions in ribosomal RNA genes show a very high degree of species specificity. This can be demonstrated by the failure of heterologous genes to be transcribed in extracts that faithfully transcribe homologous ribosomal RNA genes. In such a situation, genes from species A work in A extracts but not B extracts, whereas genes from species B work in B extracts but not those from species A. Thus a kind of molecular coevolution must occur, with genes that encode *trans*-acting transcription factors evolving in parallel with the DNA sequences these factors recognize.

Chromosomal regions containing arrays of rRNA genes are potential sites for nucleolus formation and are thus referred to as "nucleolar organizers." Although there may be more than one nucleolar organizer region, the number is rarely more than two or three per genome. It can be shown by *in situ* hybridization that most of the DNA that hybridizes to ribosomal gene probes is located in these regions; however, not all the rRNA genes are contained within the nucleolus itself. Many of the genes exist in an apparently inactive form just outside the nucleolus. Using genetic stocks of maize containing chromosomal translocations involving DNA near the nucleolus, R. Phillips and his associates at the University of Minnesota were able to show that DNA from this region could organize a nucleolus when it was transferred to another part of the genome, even though it had not done so at the original location. Thus the inactive genes are capable of functioning when given the chance, and their inactivity must reflect the operation of a system for regulating the number of genes that can be active in any one nucleus.

A similar conclusion can be drawn from experiments in wheat. Since bread wheat is a hexaploid species it is possible to make various aneuploid derivatives in which the number of nucleolar organizers, and the total number of rRNA genes, varies considerably. R. Flavell and his colleagues at the Plant Breeding Institute in Cambridge, England have shown that the total nucleolar volume, which is an index of rRNA gene activity, remains relatively constant as the number of genes is varied over a wide range. This dosage compensation effect is evidence that the activity of rRNA genes is regulated by some mechanism independent of their copy number.

Additional experiments with aneuploid wheat lines and wheat lines that contain chromosomes from related species show that different nucleoli exhibit a "dominance hierarchy." Thus, for example, the nucleolus on chromosome 1B is always larger than that on chromosome 6B in the euploid cultivar "Chinese Spring." Since the nucleolar organizer on 6B has about twice as many rRNA genes as that on 1B, it is not possible to explain dominance simply on the basis of gene number. Instead, it seems more likely that something about the genes on 1B makes them more able to compete for some limiting factor or factors required for activity. In this connection, it is striking that the dominant genes generally contain larger numbers of the spacer subrepeats mentioned previously in connection with size polymorphism. There is a suggestion that certain subrepeats function like enhancers to regulate *cis*-associated rRNA genes.

**Families of genes that code for proteins.** A number of plant and animal protein-coding genes are also found in multigene families. However, these families are much smaller than the rRNA gene families, and it is much more common for protein-coding genes to be individual units, even when they are located fairly closely together. A good example of clustered individual units is provided by the genes for the small subunit of RuBP carboxylase (rbcS genes). These genes have been extensively studied in several plant species, including petunia, pea, and duckweed (*Lemna gibba*). In all three cases there are multiple copies. *Lemna* has about thirteen copies, pea has six, and there are eight in petunia. Although the genes in any one plant are highly homologous within their protein-coding regions, they differ from one another in transcribed, but untranslated, regions located at the 5' and 3' ends of their coding sequences. By making gene-specific probes from the 5' or 3' regions or by sequencing cDNA clones and comparing their sequences to those of the genes, it is possible to show that most of the genes are actually transcribed. Therefore, the multiplicity of gene copies does not merely reflect the presence of inactive copies or "pseudogenes" (partial or defective copies), although the presence of some pseudogenes cannot be ruled out.

Genetic variation in the small subunit protein had previously been shown to be consistent with the presence of a single Mendelian locus. This could mean that (a) the multiple rbcS genes are located so close together that recombination among them is very improbable, or that (b) the variation actually reflects the activity of a modifier gene, rather than the structural gene, or that (c) only one gene is responsible for the bulk of the protein, the others being inactive in the cells that were analyzed. Recent experiments, which exploit the ability to distinguish different genes directly at the DNA level, have shown that the structural genes do actually occur quite close together in the pea genome. Differences in restriction sites in or near the gene were used as genetic markers in conventional genetic mapping experiments. This approach, called restriction fragment length polymorphism (RFLP) mapping, is illustrated in Figure 4–17. It

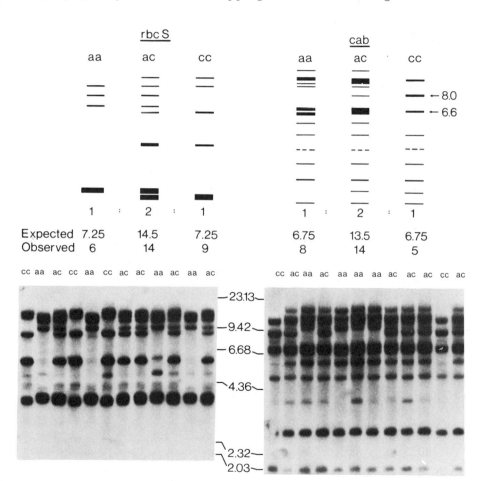

**Figure 4-17.** Restriction fragment polymorphisms of DNA containing *rbc*S sequences. The banding patterns aa and cc show allelic differences between the several genes in two strains of pea plants. F₂ progeny (bottom) show parental or heterozygous patterns only, suggesting tight linkage between the several genes. From Polans, and others, *Proc. Natl. Acad. Sci.* USA 82:2, 1985.

shows that the genes are located close together on the genetic map, where they probably constitute a single locus. In principle, different copies of the rbcS genes could still be very far apart in molecular terms; the genetic data are equally consistent with a tightly clustered pattern and also one in which the rbcS genes would be spread over several hundred thousand bases of DNA. Evidence from genomic cloning experiments, however, provides support for the tight cluster model, since it has been observed in several instances that single clones contain two different genes. The largest cluster was described by C. Dean and her colleagues at Advanced Genetic Sciences, Inc., who found five rbcS genes and a gene fragment in a set of overlapping clones defining a region of about 23 kb in the petunia genome. (One kb equals 1000 base pairs of double stranded DNA.) Taken together with the genetic mapping data for pea, these observations support the notion that members of the rbcS gene family are located close to each other in the genome.

A very similar picture has emerged for the soybean leghemoglobin genes studied by Verma and his colleagues at McGill University. In this case (Fig. 4–18), four regions contained sequences homologous to leghemoglobin mRNA; two of these contained four intact genes. Three of the four intact genes in the genome were found within a region of approximately 15 kb. All four regions containing leghemoglobin sequences also contained at least one member of a particular repeated sequence family (the s and s' elements in the figure) located near the 3' end. Multiple leghemoglobin genes may have been formed by gene duplication events—some of which could have been associated with tetraploidization—during the evolution of modern soybeans. Subsequent deletion events may then have removed most of the duplicated sequences in those regions that now contain only gene fragments.

Multigene families typically can be divided into subfamilies on the basis of the degree of homology between different groups of family members. Perhaps the best studied example of this phenomenon is the zein multigene family studied by Larkins and his colleagues at Purdue University. Zein proteins can be divided into two major classes on the basis of their molecular weights. The DNA sequences of these two classes show considerable homology, indicating that both classes probably evolved from the same ancestral gene and can be regarded as belonging to one large gene family. This large family can be divided into the two groups just mentioned, and one of the groups can be further subdivided. From an analysis of Southern blots it appears there may be at least 40-50 genes in an inbred line of maize. Although not all of these genes may be functional, the isolation of cDNA clones containing at least ten different sequences indicates that considerable heterogeneity exists even among genes that are expressed.

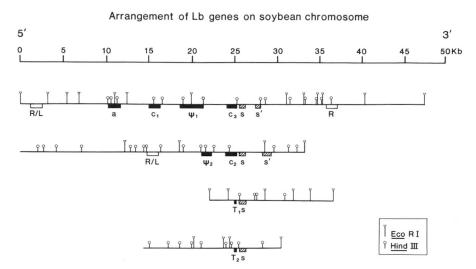

**Figure 4-18.** Chromosomal arrangement of leghemoglobin genes and related sequences in soybean. There are at least 4 loci, two of which contain only partial gene sequences. Here R and R/L are sequences (not related to leghemoglobin) that are expressed in roots or roots and leaves, respectively; s and s' are repeated sequences. From Brown, and others, *Jour. Molecular Evolution* 21:19, 1984.

Not all members of a multigene family that are expressed are necessarily expressed equally or regulated in the same way. In pea plants, for example, G. Coruzzi and her colleagues in N.-H. Chua's laboratory at Rockefeller University showed that several different rbcS genes were transcribed to yield mRNAs in light-grown plants, but that the fraction of the total rbcS mRNA contributed by a particular gene varied from one organ to another. In general, it would seem that coupling different members of a gene family to different regulatory elements might allow the "same" gene to be expressed differently in different organs or cell types, or at different stages of development. Such a system might considerably increase the range of possible developmental patterns.

# FIVE _____

# Genomes of Chloroplasts and Mitochondria

## A. CHLOROPLAST DNA

**Structure of the chloroplast genome.** The chloroplast genomes of vascular plants and most algae are quite similar in general structure and organization, especially by comparison to the wholesale variation seen in the nuclear and mitochondrial genomes. With one possible exception, all known chloroplast genomes are circular DNA molecules. The most size variation is found among green algae, where most chloroplast genomes range between about 85 and 300 kb. The genome of *Acetabularia* chloroplasts is exceptional in that it is very large (approximately 2,000 kb) and may be composed of linear rather than circular DNA molecules. However, in angiosperms, our principal concern in this book, chloroplast genomes, in all but two of over 200 examined species, are circular and range in size between 120 and 160 kb. The low end of this range is a single group of legumes, which lack one copy of a large (15–25 kb) repeated sequence characteristic of most other chloroplast genomes (discussed subsequently). Thus the great majority of angiosperm chloroplast genomes actually fall into the relatively narrow range between 135 and 160 kb.

The size of the chloroplast genome is small enough to be easily analyzed by restriction mapping techniques. One can select restriction enzymes that cut chloroplast DNA into relatively few, perhaps ten to twenty, fragments. These can be resolved from one another by agarose gel electrophoresis. By measuring the size of each fragment and adding them together, one can estimate the size of the genome. Using several different restriction enzymes and one or more of a number of different restriction mapping techniques (see technical insert at the end of this section), it is possible to produce a map showing how the various restriction enzyme fragments fit together. In all cases where restriction maps have been constructed, the restriction maps are circular—that is, the linkages between fragments define a single circular map. These mapping data do not actually prove that the genome exists physically as a circular DNA molecule since on the basis of mapping data alone one might also postulate a "circularly permuted" series of linear molecules with different end points. However, carefully prepared chloroplast DNA can be shown by electron microscopy to contain many circular molecules, which form a homogeneous class with a size consistent with the genome size estimates from restriction analysis. One sees linear molecules in these preparations, but their presence is easily explained by breakage resulting from shearing or nuclease damage during isolation. Thus, although the presence of a small fraction of linear molecules is difficult to rule out, it is generally accepted that chloroplast genomes are circular DNA molecules. Furthermore, it seems that each circle contains a single copy of the entire genome. A map of the spinach chloroplast genome (Fig. 5–1) shows the circular structure and the location of several protein-coding and ribosomal RNA genes.

Many copies of the chloroplast DNA are normally present in each cell and in each plastid as well. The number of genomes per plastid is variable. During leaf development up to 200–300 genomes per plastid may exist in very young leaf cells, but this number decreases as a more or less constant amount of DNA is apportioned among an increasing number of plastids in each cell. In a mature leaf cell one might find several hundred chloroplasts, each with as many as fifty genomes. This means a total of 5,000–10,000 copies of the chloroplast DNA in each cell. The very high effective ploidy of the chloroplast genome means that a significant

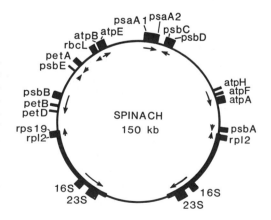

**Figure 5-1.** Map of the spinach chloroplast genome showing the large inverted repeats (heavy line) and a selection of protein-encoding and ribosomal RNA genes (solid boxes). The direction of transcription is indicated by the arrows inside the circle. See Table 5-1 for identification of genes. Reproduced with permission from J.D. Palmer, in *Annual Review of Genetics,* Volume 19, © 1985 by Annual Reviews Inc.

fraction of the total DNA in a cell—as much as 30% if the nuclear genome is small—can be of chloroplast origin.

A major structural feature of most chloroplast genomes is the large inverted repeat. One unit is composed of two sequences, usually about 20–25 kb in size, and contains the genes for chloroplast ribosomal RNA. An inverted copy occurs in most chloroplast genomes, with the exception of the genomes of certain legumes, including peas, broad beans, and alfalfa. Because few species lack the inverted repeat, and because plants more primitive than the legumes (for example, liverworts) contain the inverted repeat, it is thought that one copy was lost by a relatively recent ancestor of the legumes that lack the repeat. In higher plants the inverted repeat sequences are usually positioned in the genome so that one of the single copy regions is about four times the length of the other. Sequences in the two segments of the repeat usually are identical, even though mutations in this region lead to sequence divergence between species. Thus, some kind of copy correction mechanism must operate to maintain the identity of the two repeats in the face of mutational divergence.

Intramolecular recombination can occur at the large inverted repeats, leading to an isomerization (or "flipping") between the two possible relative orientations of the single copy regions. Recombination between the inverted repeats of different molecules occasionally produces head-to-head circular dimeric chromosomes as shown in Figure 5–2.

**Gene content and arrangement.**  Chloroplast DNA encodes a complete set of ribosomal RNAs and tRNAs as well as many proteins. The ribosomal RNA genes are contained in the inverted repeat and thus are present in two copies per genome. They are organized into an operon which has many similarities to bacterial rRNA operons, as will be discussed subsequently. About forty tRNAs, including acceptors for all twenty amino acids, have been identified and mapped in the chloroplast genomes of several species of plants. Although tRNA genes are arranged in clusters in the *Euglena* chloroplast genome, they are dispersed throughout the genome of higher plants. An exception is the presence of two tRNA genes in the transcribed spacer on the 5' end of the rRNA operon. (See the discussion which follows.)

The approximately 120–140 kb of sequence complexity (ignoring repetitions) in the chloroplast genome is theoretically sufficient to code for roughly 100 average size proteins. Early estimates were made from experiments on protein synthesis in isolated plastids. If only plastid-coded proteins are synthesized in light-dependent reactions in isolated chloroplasts, and if one can count all the proteins so synthesized, it is then possible to estimate the number of protein-coding genes. The best of these experiments revealed about 80 different soluble (not membrane) peptides by two-dimensional PAGE. Some of the peptides were probably artifacts

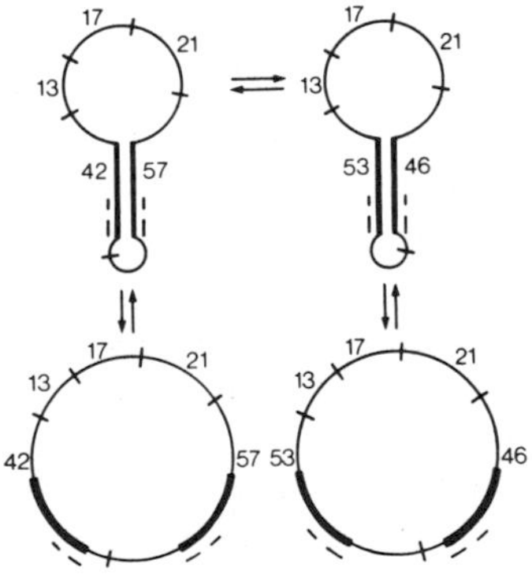

**Figure 5-2.** "Flip-flop" isomerization of chloroplast DNA. The figure illustrates two isomeric forms that are distinguished by a reversal of the polarity of the small single copy region relative to the large single copy region (note the position of the restriction site in the small single copy region). Heavy solid lines indicate the inverted repeats. Ribosomal genes in the inverted repeats are diagrammed as lines outside the circle. In the top portion of the figure the chloroplast genome is shown in a "dumbell" configuration to emphasize how recombination might occur between the two halves of the inverted repeat. The figure also shows how restriction enzyme analysis can be used to distinguish the two forms. An enzyme must be found that cuts once, asymmetrically, within one of the single copy regions and does not cut within the inverted repeat sequences. Restriction sites and fragment sizes are illustrated for *Sal* I and *Phaseolus* chloroplast DNA. Note the difference between the two isomeric forms in the sizes of the two largest fragments. Finding all four of these fragments in a restriction digest would indicate the presence of both isomers in the preparation. From J.D. Palmer, in *Molecular and Evolutionary Genetics*, R.J. MacIntyre, ed., 1985.

produced by premature chain termination in the *in vitro* conditions. On the other hand, certain membrane proteins known to be encoded by the chloroplast genome would not have been detected. All things considered, an estimate of 80–100 proteins was considered reasonable.

Complete DNA sequences for entire chloroplasts genomes are now available, and it is possible to determine the number of genes more precisely. Computers are used to search the data for sequences with the properties of genes, for example, "open reading frames" which contain a series of amino acid codons beginning with an initiation codon and ending with a termination codon and sequences that are homologous to known tRNA genes. From such work, first carried out by two groups in Japan, it is possible to predict the presence of more than 120 genes. About thirty of these are tRNA genes and another four encode ribosomal RNAs. There are about eighty-five protein coding genes, a number that is in remarkably good agreement with the earlier estimates based on two-dimensional PAGE analysis. A majority of these encode proteins that have been identified in chloroplasts or that are similar to proteins identified in other organisms, such as ribosomal proteins and RNA polymerase subunits.

Associating bona fide chloroplast proteins with particular chloroplast genes requires steps beyond DNA sequence analysis. The most common approach involves purifying a protein of interest and making an antibody to it. One can then determine the location of genes coding for the protein by one of two methods. In the first method extracts from *E. coli* cells are used to transcribe isolated restriction fragments of chloroplast DNA. The resulting RNA is then translated into protein, either in the same reaction or in a second reaction with a different extract. The proteins produced are then tested with the antibody to see which restriction fragment programs the synthesis of the protein of interest. This method takes advantage of the fact that the structures of many chloroplast genes, including their promoters and regulatory signals, are quite similar to those of bacterial genes. However this method does not always work because, for example, the gene may not be entirely contained on any single restriction fragment. In such cases it is necessary to isolate the actual chloroplast mRNAs transcribed from the candidate restriction fragment. This can be done by hybridization, using DNA immobilized on a solid support of some kind. After the hybridization reaction, unhybridized RNAs are washed away and the hybridized RNA is isolated. This RNA can then be tested in an *in vitro* translation system and the products tested with antibody just as in the first method.

Once a gene has been mapped on the chloroplast genome of one species it can usually be located on the maps of other chloroplast genomes by relatively simple hybridization procedures. One obtains a clone of a DNA fragment from the gene (preferably an internal fragment, without flanking sequences) and hybridizes this to Southern blots made of restriction digests of chloroplast DNA from the second species. From the results, one can determine which fragments of the second genome contain sequences homologous to the gene. This approach is greatly facilitated by the fact that the DNA sequences of chloroplast genes generally are highly conserved between species. It is usually quite easy to cross-hybridize homologous genes from monocots with dicots and even algae with those of higher plants.

Table 5–1 lists some of the protein-coding genes which have been mapped on the chloroplast genome in one or more species of higher plants. The positions of many of these are indicated in Figure 5–1.

All green plants for which we have information contain essentially the same set of genes in their chloroplast genomes. The reason for this similarity is not immediately obvious. One possible explanation is based

**TABLE 5-1    Chloroplast Genes which have been mapped on Chloroplast DNA.**

Ribosomal RNAs

Ribosomal RNA operons are designated rrnA, B, C, and so forth. depending on the number of gene sets in the particular genome. In most chloroplasts there is one operon in each segment of the inverted repeat. These are normally identical in sequence. By convention, rrnA is on the right hand side of the chloroplast DNA restriction map when it is drawn with the large single copy region at the top. The rbcL gene is on the left side. Each operon normally includes genes for 16S rRNA, 23S rRNA, 5S rRNA, and 4.5S rRNA.

Transfer RNAs

tRNA genes are designated "trn" to indicate transfer RNA, followed by the single letter amino acid code indicating the amino acid accepted by the tRNA encoded by the gene. Where there is more than one gene for a particular amino acid, the isoaccepting species can be indicated either with sequential numbers or by giving the anticodon. About 40 tRNA genes are known to exist in the chloroplast genome.    Examples:

  trnF—gene for tRNA(Phe)
  trnC—gene for tRNA(Cys)
  trnL1 (or trnL-UAA)—gene for tRNA(Leu)1
  trnL2 (or trnL-CAA)—gene for tRNA(Leu)2

Ribosomal Proteins

  rps 4—ribosomal protein homologous to *E. coli* ribosomal protein S4
  rps 19—ribosomal protein homologous to *E. coli* ribosomal protein S19
  rpl 2—ribosomal protein homologous to *E. coli* ribosomal protein L2

Photosystem I Proteins

  psaA1—P700 chlorophyll a apoprotein
  psaA2—P700 chlorophyll a apoprotein

Photosystem II Proteins

  psbA—"32 kilodalton" quinone-binding polypeptide. Also known as "photogene 32" and "Qb protein", it contains the binding site for atrazine-type herbicides.
  psbB—51 kilodalton chlorophyll a-binding polypeptide or P680 apoprotein
  psbC—44 kilodalton chlorophyll a-binding polypeptide
  psbD—"D2" protein
  psbE—cytochrome *b559*

Photosynthetic Electron Transport Proteins

  petA—cytochrome *f*
  petB—cytochrome *b6*
  petD—subunit 4 of the cytochrome *b6/f* complex

Proteins of the ATP Synthase Complex

  atpA—CF1 alpha subunit
  atbB—CF1 beta subunit
  atpE—CF1 epsilon subunit
  atpH—CF0 subunit III, DCCD-binding proteolipid of the ATPase complex (proton translocating subunit)

Carbon Fixation Enzymes

  rbcL—ribulose bisphosphate carboxylase, large subunit

Other Stromal Polypeptides

  tufA—translational elongation factor Tu

on the widely accepted notion that chloroplasts arose from cyanobacterial endosymbionts. Since cyanobacterial genomes are some 20–30 times larger than those of chloroplasts, a dramatic reduction in the plastid genome must have occurred after the endosymbiotic association was established. If this reduction occurred quite soon after the original endosymbiosis and prior to the divergence of lineages leading to different groups of green plants, the similarity of green plant chloroplast genomes might reflect common ancestry. On the other hand, one might postulate that certain genes are retained in the chloroplast genome for reasons of function. For example, genes for ribosomal and transfer RNAs may have been retained in the chloroplast because it is difficult or impossible to transport RNAs across the chloroplast membrane. The same may be true for certain chloroplast proteins. Lawrence Bogorad of Harvard University has suggested that natural selection might act to keep certain genes in the plastid genome if their products were required to "lock in" or promote the accumulation of certain other proteins originally synthesized in the cytoplasm. This might apply, especially, to multicomponent protein complexes composed of chloroplast and nuclear-coded proteins such as the enzyme ribulose bisphosphate carboxylase (rubisco), the chloroplast ribosomes, the ATP synthase and cytochrome $b6/f$ complexes, and both the photosynthetic reaction centers. A related hypothesis suggests that chloroplast-coded proteins might be important in regulating the level of such multicomponent complexes. This would insure that control of plastid functions remains in the plastid genome. And it would help to explain the remarkable fact that so many multisubunit complexes include components coded by both chloroplast and nuclear genes.

Although we do not yet have much data on chloroplast DNA in the chromophytic and the rhodophytic algae, it is clear that these genomes do encode a number of the same genes found in higher plant chloroplasts. In some cases they have also been shown to contain genes not present in the plastid genomes of higher plants. One particularly interesting example is the presence of rbcS, the gene for the small subunit of rubisco, on the chloroplast genome of the algae *Cyanophora* and *Olisthodiscus*. In both cases the rbcS gene is tightly linked to rbcL, the large subunit gene, and the two genes may be transcribed onto one RNA. Another example involves the phycobilisomes, complex light harvesting structures present in cyanobacteria and certain nongreen algae. In *Cyanophora,* genes for several phycobilisome proteins are encoded in the plastid DNA.

The arrangement of genes on the chloroplast genome can vary widely between distantly related organisms, such as higher plants and algae, and there are cases in which gene order varies dramatically even between closely related species. These cases are the exception rather than the rule, however. In general, changes in gene order are extremely rare among vascular plants. Studies in which cloned fragments of chloroplast

DNA from one species are hybridized to Southern blots of restriction fragments from another species show that twenty-four families of angiosperms (of the thirty examined) have essentially the same gene order as the spinach genome (Fig. 5–1). The spinach gene order is also found in the gymnosperm *Ginkgo biloba* and the fern *Osmunda cinnamomea*. The gene order of spinach may be the ancestral gene order for all vascular plants. Where differences in gene order exist, they can usually be explained by postulating a small number of inversion events. For example, the mung bean chloroplast genome differs from that of spinach in that a large segment of the large single copy region is present in an inverted orientation. The end points of this inversion are such that the rbcL gene, which is 50 kb away from the psbA gene in spinach, is only 5 kb from the psbA gene in mung bean. Within the inverted segment the gene order is preserved—that is, there have apparently been no other major rearrangments. This inversion seems to have occurred relatively early in the evolution of the legume family, since it is found in all tested genera of the subfamily Papilionoideae, but not in any of the other legume species examined. Several other examples of large inversions have been encountered, but, clearly, they are rare events in chloroplast genome evolution.

Much more extensive alterations in the plastid genome have occurred during the evolution of another group of legumes—the pea, broad bean, and clover group (Fig. 5–3). Interestingly, this group consists of species whose chloroplast DNA does not contain the large inverted repeat found in all other higher plants and green algae. It appears that loss of the inverted repeat occurred early in the evolution of this particular lineage and may have somehow led to an increase in the frequency of rearrangements in subsequent evolution. Although the details of this evolutionary process are not understood, the extensive rearrangments in the genomes of pea, broad bean, and clover contrast strikingly with the very high degree of conservation seen in essentially all other vascular plant chloroplast genomes.

**Ribosomal RNA operons.**    Chloroplast ribosomal RNA genes are arranged in an operon very similar to the ribosomal RNA operon of bacteria. Figure 5–4 shows the structures of several chloroplast operons and those of *E. coli* and *B. subtilis*. The bacterial gene order, 16S—23S—5S, is preserved in chloroplast operons. The "addition" of a gene for 4.5S rRNA may be a rearrangement of DNA rather than an insertion since this RNA is homologous to the 3' end of the bacterial 23S RNA and can be regarded as the structural equivalent of this portion of the molecule. An additional point of similarity is the location of tRNA genes for isoleucine and alanine in the spacer between the 16S and 23S rRNA genes. The same two tRNA genes can be found at corresponding positions in several of the rRNA operons of both *E. coli* and *B. subtilis*. However, in most

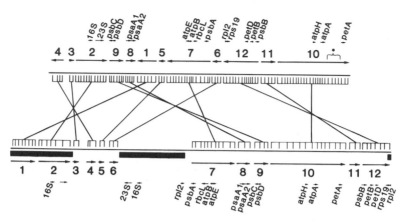

**Figure 5-3.** Rearrangements of sequences in the pea chloroplast genome relative to that of mung bean. A linearized version of the mung bean chloroplast genome is shown on the bottom of the figure. The large inverted repeats are indicated by the heavy black lines, while the positions and directions of transcription of selected genes are shown by the small arrows above the gene names. See Table 5-1 for identification of genes. Restriction sites demarcating the fragments used in interspecific hybridization experiments are indicated by the vertical lines. Lines between the mung bean genome and the pea genome at the top of the figure indicate patterns of cross-hybridization between the mung bean fragments and fragments of the pea genome. It is apparent from the figure that these genomes may be subdivided into "rearrangement units," which have been indicated by the large numbered arrows. See text for further discussion. Reproduced with permission from J.D. Palmer, *Annual Review of Genetics*, Volume 19, © 1985 by Annual Reviews Inc. Photograph courtesy of Dr. Palmer.

chloroplasts these two tRNA genes contain large introns which are not present in bacteria.

The homology between plastid and bacterial rRNA operons is also apparent from DNA sequence. Though intergenic sequences and intron sequences diverge rapidly, and large differences are seen even between different species of higher plants, there is a high level of homology between corresponding structural genes (exons) of different organisms. The sequences of bacterial and higher-plant plastid rRNA genes show 60% to 80% similarity. A strong homology between plant and bacterial sequences at the 3′ end of the 16S gene, which in bacteria is known to be involved in binding the ribosome to the mRNA at the start of translation, provides evidence that this process may be similar in plastids. Supporting this notion is the fact that chloroplast mRNAs have often been found to contain sequences similar to the ribosome binding sites of bacterial mRNAs.

Transcription of the plastid rRNA operon probably produces a single primary transcript containing the 16S, 23S, 4.5S, and 5S rRNAs, along with spacer tRNAs. This is then processed by a series of endonuclease

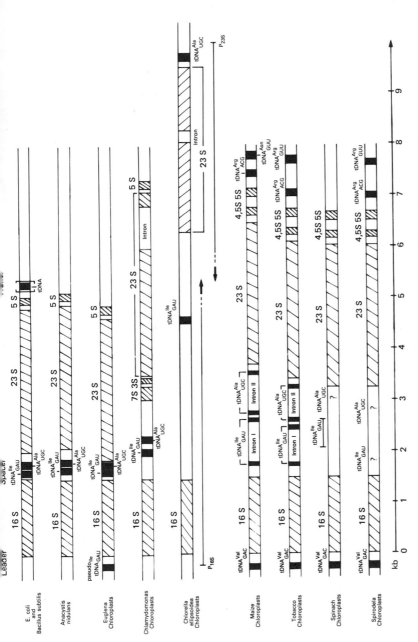

**Figure 5-4.** Comparison of rRNA operons in chloroplast and bacterial genomes. Genes coding for the various ribosomal and transfer RNAs in the region are shown together with promoter (P) sites in one case. Reproduced with permission from H. Kossel, and others, *Molecular Form and Function of the Plant Genome*, pp. 183-198, L. van Vloten-Doting, and others, eds., 1985. Diagram kindly provided by Dr. H. Kossel.

cleavages to produce the mature RNA products. However, details of this scheme are difficult to define precisely, since the processing is so rapid that the original transcript probably never accumulates to detectable levels.

**tRNA genes.**    In *Euglena*, tRNA genes are organized into operons which are transcribed to form precursor RNAs containing several tRNA sequences. In higher plants, tRNA genes are typically dispersed throughout the genome and transcribed individually. Interestingly, although *Euglena* tRNA genes lack introns, about one-fourth of the sequenced higher plant tRNA genes have them. Thus, in both cases tRNA primary transcripts must be processed to produce mature tRNAs. In one case the processing removes intergenic sequences, whereas in the other it removes introns. Intron removal from most tRNA gene transcripts probably proceeds by a mechanism different from that used to remove introns from the transcripts of nuclear genes or protein-coding genes in the chloroplast. Instead of relying on conserved sequences at exon/intron boundaries, splicing of tRNA transcripts seems to depend on the secondary structure of the intron in a manner reminiscent of the splicing of certain ribosomal RNAs.

**Protein genes.**    Whether or not most chloroplast protein genes are transcribed as parts of operons remains to be determined. It is known that rbcS and psbA produce very abundant monocistronic transcripts, but these genes may be exceptions to the general rule.   Transcript mapping studies indicate that the situation in other regions of the genome may be much more complex. When Northern blots of electrophoretically separated RNA are probed by hybridization with small cloned fragments of chloroplast DNA, one often sees a large number of RNA bands with homology to the probe. Examples are shown in Figure 5–5. Some of the RNA molecules visualized in this way are quite large (for example, 4–8 kb), much larger than any one gene, and often many times the length of the probe. In some cases it has been shown that most of the RNAs in a series of bands come from the same strand of the DNA. Since in angiosperms most chloroplast protein-coding genes do not contain introns, this multiplicity of RNAs must reflect the use of multiple initiation or termination sites and/or the processing of a long primary transcript. In both cases the initial transcripts are polycistronic, and the production of mature, translatable mRNA must involve processing steps. Many of the intermediate sized RNA bands may be processing intermediates of various types.

These observations of polycistronic transcripts are surprising since chloroplast protein genes (in contrast to rRNA genes) generally are not organized into the prokaryotic operon pattern in which functionally re-

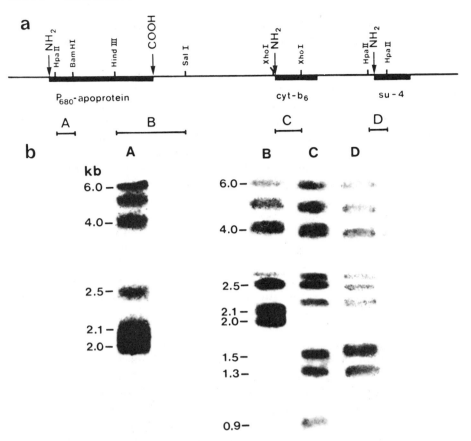

**Figure 5-5.** Complex transcript patterns for chloroplast genes. The top part of the figure (a) is a map of a segment of spinach chloroplast DNA. Probes from the regions designated A,B,C, or D were hybridized to Northern blots of electrophoretically separated RNA. Autoradiographs of the resulting hybridization patterns are shown in the bottom part of the figure (b). Lengths of the RNAs (in kb) to which these probes hybridize are indicated to the left of the autoradiographs. The largest probe fragment (B) is 0.96 kb in length, while the others range in size from 160 to 300 nucleotides. From Morris and Hermann, *Nucleic Acid Res.* 12:2837, 1984.

lated genes are closely linked. It is presumed that such operons existed in the ancestral endosymbionts that gave rise to chloroplasts, but they were lost when plastid genes were transferred to the nucleus or rearranged during evolution of the chloroplast genome.

Some remnants of a prokaryotic operon structure can be discerned in chloroplast genomes, however. One case involves genes for the thylakoid membrane ATPase complex. In *E. coli* ATPase genes are part of a well-defined operon. Some remnants of such a structure may be imagined in the plastid genome where atpB and atpE are very close together. In fact, they actually overlap in many chloroplast genomes, with the sequence ATGA containing both the first codon of the atpE coding sequence (ATG) and the translation stop codon of atpB (TGA). The genes atpA and atpH are also found fairly close together (within 2 kb), but they are located far away from atpB and atpE. (See Figure 5–1.) Genes for other components of the ATPase complex are not found in the chloroplast DNA at all but are located in the nucleus. Examples of functionally related genes scattered in different locations in the plastid genome include those for chloroplast ribosomal proteins, proteins of the cytochrome *b6/f* complex, and proteins of the photosynthetic reaction center. Additional components of some of these same complexes, notably a large number of ribosomal proteins, are encoded within the nucleus.

Relatively few protein genes in angiosperm chloroplasts have introns. Only six introns were detected in an electron microscopic analysis by B. Koller and H. Delius of *Vicia faba* chloroplast DNA-RNA hybrids. They were detected under conditions in which at least fifty introns (accounting for over 20%) in the *Euglena* chloroplast genome could be shown. Such a large difference between *Euglena* and higher plants is not necessarily surprising since their chloroplast genomes are thought to have arisen from separate endosymbiotic events. Surprisingly, in both *Euglena* and higher plants, where intron-containing chloroplast genes have been sequenced, the genes appear to have similar intron boundary sequences. This observation indicates that the mechanisms for intron excision and splicing for chloroplast protein genes may be similar to those for nuclear mRNAs in that conserved intron boundary sequences seem to direct the splicing process. This mechanism differs from intron processing in tRNA genes; therefore, two different splicing mechanisms probably exist in the chloroplast.

**Chloroplast promoter sequences.**   Searches for conserved promoter-like sequences upstream from chloroplast genes reveal elements with considerable homology to bacterial promoter sequences. The *E. coli* "consensus" promoter sequence contains two conserved regions, one normally found about thirty-five nucleotides and the other about ten nucleotides upstream from the start of transcription and referred to as the "-35" and "-10" elements, respectively:

5'...TTGACA/T.....(16–18 nucleotides).....TATG/AAT...3'

The chloroplast consensus sequence described by L. Bogorad and his colleagues at Harvard University is as follows:

5'...A/gTTG/cA/cNa/t ....(15–20 nucleotides)....TA/tA/tG/aAT...3'

(In the preceding line, lower case letters represent less frequent alternative bases.)

In several cases where the start of the RNA transcript has been identified by S1 nuclease protection experiments, it occurs within eight nucleotides of the proximal promoter element.

The conservation of these sequence elements and their homology with bacterial promoters support, but do not prove, the notion that they function as chloroplast gene promoters. Proof requires the direct demonstration that removing or changing these sequences actually affects promoter activity. One way of obtaining such proof is to test altered genes in an *in vitro* transcription system, as performed by W. Gruissem at the University of California at Berkeley. An example of such an experiment is shown in Figure 5–6. The approach in this case was to make progressive deletions of sequences 5' to the gene, moving gradually closer to the start of transcription. Each deletion mutant was characterized by DNA sequencing and then tested for its ability to support accurate transcription. As illustrated, deletions of sequences further upstream than position -85 had little effect on the production of transcript, but there was a rapid drop in transcription caused by deletions between -80 and -75. This region contains the sequence TTGCTTA, the first three nucleotides of which are homologous to the *E. coli* -35 consensus sequence mentioned previously. There is also a TATAAT sequence between -54 and -59 which is fully homologous to the *E. coli* -10 consensus sequence. Since transcription was already inhibited by the removal of the sequences further upstream, little effect was seen in this experiment when the TATAAT sequence was deleted. Data from experiments with other gene modifications establish its significance and also show that the seventeen base pair region between it and the upstream sequence is critical because of the spacing it provides rather than the particular base sequence it contains.

Much further work remains to be done to characterize the promoter sequences of different chloroplast genes, to assess the variance from gene to gene, and to evaluate the consequences any such variation may have on gene expression. In addition to promoter sequences, we will be concerned with sequences which may be involved in termination or attenuation. Combined with studies of the enzymes involved in transcription and RNA processing, such work will substantially improve our understanding of chloroplast gene expression.

## Technical insert: Restriction Mapping

Restriction mapping is fundamental to most techniques of DNA analysis. By taking advantage of the ability of restriction endonucleases to recognize specific oligonucleotide sequences, it is possible to locate a set of discrete landmarks in a segment of DNA. Once established, such a "restriction map" can be extremely useful in attempts to relate structure and function in the genome. Organelle DNAs such as the chloroplast genome and mitochon-

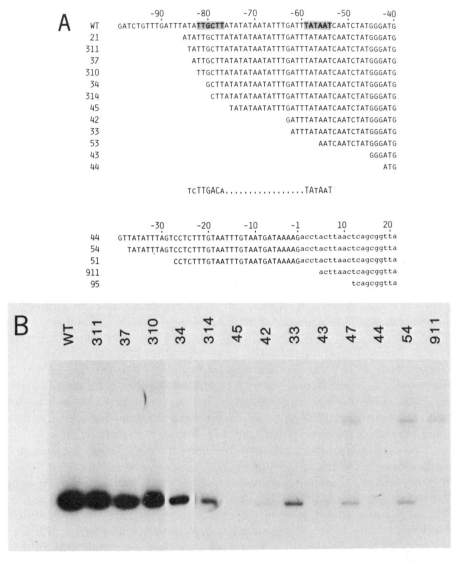

**Figure 5-6.** Analysis of the trnM2 gene promoter by deletion mutagenesis. Mutants were constructed by deleting portions of the 5′ flanking sequences as described in the text. The upper part of the figure shows the sequences of the wild type gene and several genes modified in the region upstream of position -40 (40 nucleotides upstream from the transcription start). Consensus "-35" and "-10" sequences are shown below the gene sequences. Transcriptional activity was measured *in vitro* as the amount of radioactivity in properly initiated transcripts on gels like that shown in the lower part of the figure. Clone numbers are indicated above each lane, and the amount of transcript (as a percentage of that produced by the wild type gene) is shown below each lane. From Gruissem and Zurawski, *EMBO Jour.* 4:1637-1644, 1985. Photograph courtesy of Dr. Gruissem.

drial genomes discussed in this chapter are particularly suitable for analysis by restriction mapping because they are small enough in size to permit construction of complete maps by reasonably simple procedures.

We will discuss the multiple digest method and the hybridization method of restriction mapping. The multiple digest method involves determining the sizes of fragments produced by digestion with two or more restriction enzymes used singly or in combination. One fits the various fragments together to produce a map in much the same way that one might fit pieces of a jigsaw puzzle together. For example, in Figure 5–7 we list the fragments produced by the digests performed in order to map the sites for

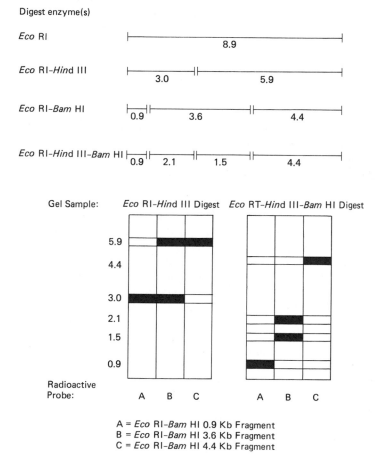

**Figure 5-7.** Examples of data that might be used to map a DNA segment. Top: a hypothetical segment, e.g., an *Eco*R I fragment of chloroplast or mitochondrial DNA cloned in a plasmid vector. Middle: fragments released by digestion with *Eco*R I plus *Hin*D III, *Bam*H I, or the combination. Bottom: patterns generated by hybridizing labeled probes with Southern blots. Dark areas represent locations of label.

*Hin*d III and *Bam*H I within a hypothetical 8.9 kb *Eco*R I fragment. *Hin*d III cuts this fragment into two pieces of 3.0 and 5.9 kb so that there is a *Hin*d III site 3.0 kb from one of the *Eco*R I ends. When the fragment is cut not only by *Hin*d III but also by *Bam*H I, neither one of the *Hin*d III fragments remains intact. From the sizes of the triple digest products we can infer that the 3.0 kb *Eco*R I-*Hin*d III fragment is cut by *Bam*H I to produce the 0.9 and 2.1 kb pieces, whereas the 5.9 kb fragment gives rise to the 4.4 and 1.5 kb segments. Including the *Eco*R I-*Bam*H I fragments in the analysis leads to the conclusion that one *Bam*H I site must be 0.9 kb from an *Eco*R I site, and another must be 4.4 kb from the other *Eco*R I site. The 4.4 kb fragment could only come from the 5.9 kb *Eco*R I-*Hin*d III fragment. The remaining 3.6 kb are accounted for by the 3.6 kb fragment in the *Eco*R I-*Bam*H I digest. The map resulting from these considerations is shown at the top of Figure 5–7.

The hybridization method is applicable to considerably more complex situations and can often overcome ambiguities introduced by fragments that are similar in size. Therefore, although it involves more complex procedures it is often the method of choice. The crucial feature of this technique is the preparation of radioactive restriction fragments (probes) from a digest prepared using one enzyme. The separated fragments are hybridized to Southern blots of the same DNA digested with a second enzyme. Assuming that no repeated sequences are present, it is easy to see when two fragments produced by one enzyme contain sequences found on one fragment produced by the other enzyme. When this is true, the two fragments must be adjacent. Taking the simple example in Figure 5–7, we prepare probes separately from each of the three *Eco*R I-*Bam*H I fragments. We then hybridize each of them to the *Eco*R I-*Hin*d III and *Eco*R I-*Bam*H I-*Hin*d III fragments. The hybridization patterns are indicated in the bottom of the figure. Hybridization of the 3.6 kb *Eco*R I-*Bam*H I probe fragment to the *Eco*R I-*Bam*H I-*Hin*d III blot establishes linkage between the 2.1 and 1.5 kb fragments. Hybridization to the *Eco*R I-*Hin*d III blots shows the linkages between the 0.9 and 2.1 kb fragments and the 1.5 and 4.4 kb fragments. Although the hybridization approach usually involves more work than is necessary to make a simple map like that shown here, it has proved exceptionally powerful for mapping much larger numbers of fragments quite rapidly—as in the case of entire chloroplast genomes.

A variant of the hybridization procedure can be applied on an even larger scale by using cloned DNA fragments instead of genomic DNA and by looking for homologies between sequences in different clones to provide linkage information. This type of mapping is often called "chromosome walking," since one proceeds by using the sequences at one end of a given clone to "step" to the next (overlapping) clone. One repeats the process as often as necessary to cover the desired region of DNA. With careful attention to the choice of probes and the criteria of homology, it is possible to use this technique to make maps of genomes containing a certain number of repeated sequences. In principle, there is no limit to the size of the genome that can be mapped in this way. Since larger regions of DNA can be mapped more quickly if one uses larger clones, chromosome walking is often carried out with bacteriophage clones (see technical insert on cloning in Chapter 4, Section 2), where inserts may be 15 kb or more in length. Or "cosmids" can be used in which inserts can be about 40 kb. Cosmids are engineered hybrids between plasmids and bacteriophage. They have the plasmid origin of replication and the *cos* sites from bacteriophage *lambda*. They replicate in *E. coli* as a plasmid, but the *cos* sequence permits cosmid DNA to be packaged into phage heads *in vitro* to yield high transformation efficiencies. Since the replication of *cos* functions requires very little DNA in comparison to a normal phage genome, there is room in the phage particles for very large inserts. Chromosome walking with cosmid clones was used to produce a restriction map of the 570 kb maize mitochondrial genome and some of its recombinant forms, as described subsequently.


## B. MITOCHONDRIAL GENOME

**General structure.**    The plant mitochondrial genome has long presented a puzzle to molecular biologists. Even the smallest is more than 200 kilobases in size, more than 10 times the size of animal mitochondrial genomes (15–18 kb) and several times the size of mitochondrial genomes in fungi (18–78 kb) or protists (15–47 kb). The largest plant mitochondrial genome studied so far is about 2,400 kb, or more than half the size of the entire *E. coli* genome (4,500 kb). The large sizes of these genomes present several problems. On the one hand, simply determining the physical structure of such large genomes is a major challenge, especially since recombination events are known to produce several different molecular configurations. On the other hand, the large size per se and the dramatic variation of such genomes demands an explanation. We know of few genes in plant mitochondrial genomes which are not also present in the mitochondria of yeast or animal cells. And the few additional genes we do know about do not begin to account for the additional DNA in even the smallest of plant mitochondrial genomes. Both animal and plant mitochondria encode their own ribosomal and transfer RNAs. The number of proteins encoded in plant mitochondrial DNA is probably not much higher than the number encoded in mammalian mitochondria. Experiments in the laboratory of Chris Leaver in Edinburgh have shown that most of the protein synthesis in isolated maize mitochondria can be accounted for by some 18-20 polypeptides. Although there is always a possibility that more proteins might be synthesized in mitochondria *in vivo* than *in vitro*, it seems likely that most of the mitochondrial genome is noncoding DNA.

Changes in the size of the mitochondrial genome seem to occur quite rapidly in plant evolution since closely related species sometimes have quite different mitochondrial genome sizes. The best example of this phenomenon comes from studies in the laboratory of Arnold Bendich at the University of Washington. Combining measurements of mitochondrial DNA renaturation kinetics with an analysis of restriction profiles for the same DNAs, Bendich and his colleagues B. Ward and R. Anderson estimated the minimal size ("complexity") of mitochondrial DNA for several species in the family Cucurbitaceae. Their estimates, along with similar estimates for several other plant mitochondrial genomes, are summarized in Table 5–2. Within the Cucurbitaceae the size of the mitochondrial genome can vary by at least 10-fold. This variation cannot be explained by postulating rapid changes in the amount of repetitive DNA since little repeated DNA (less than 10%) was found in any of the genomes. Therefore, the mitochondrial genome is like the chloroplast genome in being composed principally of single copy DNA. But it is like the nuclear genome in its highly variable size and its content of "excess" DNA which has no known function.

In several cases, including *Brassica,* maize, and wheat, restriction maps have been constructed for mitochondrial genomes. Although the smaller genomes (for example, *Brassica*) can be mapped by procedures such as those used for chloroplast DNA, the larger genomes require different techniques. The most successful approach has been to clone large fragments of mitochondrial DNA and then use a combination of hybridization and restriction mapping to identify overlapping clones and es-

TABLE 5-2   **Sizes of some higher plant mitochondrial genomes.**

Where mapping data are available (*Brassica*, maize, wheat), the sizes given in the table refer to the size of the largest circular linkage group. In other cases they are complexity estimates derived from reassociation kinetics and/or restriction profiles of total mitochondrial DNA.

| | |
|---|---|
| Cucurbits | |
| Watermelon | 320 |
| Zucchini | 900 |
| Cucumber | 1,600 |
| Muskmelon | 2,600 |
| Other Dicots | |
| *Oenothera* | 200 |
| *Brassica* | 215 |
| Pea | 430 |
| Mung Bean | 400 |
| Pokeweed | 330 |
| Spinach | 300 |
| *Atriplex rosea* | 290 |
| *Atriplex halimus* | 260 |
| Monocots | |
| Maize | 570 |
| Wheat | 430 |

tablish linkage groups. This procedure is called "chromosome walking" (mentioned previously). With such a procedure it has been possible to show that most and possibly all of the mitochondrial DNA can be described as one large circular linkage group.

We have already pointed out (Chapter 5, Section A) that a circular linkage map is an abstraction and does not prove that DNA is really circular. In contrast to chloroplast DNA, in which circular molecules can be seen in an electron microscope, the physical form of the mitochondrial genome is still not well understood. Many investigators have reported the presence of circular DNAs in plant mitochondria, but these are small in relation to the total genome size. Some are mitochondrial plasmids (see following discussion), which are not considered to be part of the main genome; whereas others clearly contain genomic sequences. In some cases most of the DNA sequences in the mitochondrion can be found in the collection of circles, although individual circles are often much smaller than the size of the genome. Cultured tobacco cells provide a good case in point. These cells are an excellent source of the relatively small mitochondrial DNA circles which can be purified on density gradients and analyzed with restriction enzymes. Although none of the circles comes close to the size of the genome as a whole, the restriction profile of the DNA in the circular fraction is identical to that of total mitochondrial DNA from either cultured cells or intact plants. At least in cultured tobacco cells, therefore, it seems that the entire mitochondrial genome can exist as a population of subgenomic circles. Mitochondrial DNA consisting predominantly of small circles may be peculiar to cultured plant cells since it is often difficult to isolate any circular DNA at all (with the exception of plasmids) from the mitochondria of mature plants. In these cases the physical form of the DNA remains unknown. Although it is logical to suppose that larger circles are present *in vivo*, technical difficulties make it quite difficult to test this hypothesis.

The complex and variable pattern of mitochondrial DNA organization was difficult to rationalize until complete restriction maps became available. The relatively small *Brassica* mitochondrial genome was mapped by J. Palmer and C. Shields at the Carnegie Institution's Department of Plant Biology, using relatively simple mapping techniques developed originally for chloroplast DNA. Meanwhile, D. Lonsdale and his colleagues at the Plant Breeding Institute in Cambridge, England had been using chromosome walking techniques to map the much larger maize mitochondrial genome. The two groups reported their findings nearly simultaneously. In both cases it was discovered that the genome contained a set of repeat sequences which could be found associated with different permutations of flanking sequences, defining a set of substoichiometric restriction fragments. These data are most consistent with a model in which site-specific recombination occurs at the repeated sequences, generating a series of subgenomic circular molecules that are in conformational equilibrium with each other and with a "master circle." Figure 5–8 illustrates how recombination might produce various circular forms. By postulating changes in the recombination frequency or a differential replication of different sized molecules under different conditions, the model can also explain the differences between intact plants and cultured cells in the proportion of circular DNAs which can be isolated.

Figure 5–9 shows a number of circular forms which can be inferred to exist in maize mitochondrial DNA on the basis of Lonsdale's mapping data. Although extremely complex, the size distribution inferred from these data agrees, at least for the smaller circles, with previous electron microscopic measurements of circular DNA molecules carried out by C. S. Levings and his colleagues at North Carolina State University. The larger circles were not seen in the electron microscope, presumably because of their great sensitivity to breakage during isolation. Therefore, the available evidence from two independent approaches is consistent with a complex organization of mitochondrial DNA into a "master circle" and a number of smaller circles that are related to each other and to the master circle by recombination at specific repeated sequences.

Recombination in mitochondrial DNA has also been shown to occur

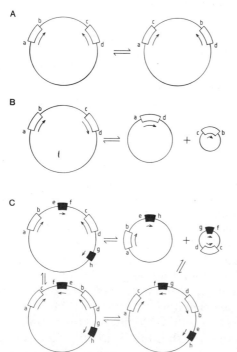

**Figure 5-8.** Possible arrangements for repeated sequences in the mitochondrial genome and their consequences in terms of recombination. In A, recombination between a single pair of inverted repeats leads to "flip-flop" as described for chloroplast DNA in Figure 5–2. In B, recombination between a single pair of direct repeats leads to "loop out" and the formation of two circular products. In C, the above two configurations are combined in a single molecule. Recombinations can then lead to four different genome configurations, three of which are single circles and one of which is a two circle composite. From Lonsdale, *Plant Molecular Biology* 3:201, 1984.

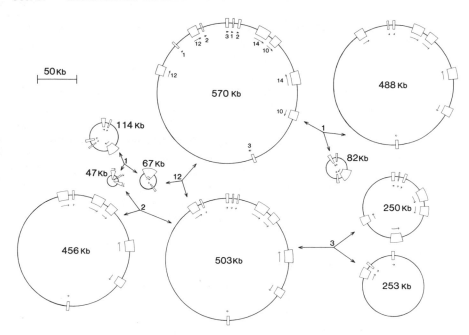

**Figure 5-9.** Circular forms of the maize mitochondrial genome. A "master circle" of 570 kb, which contains all the sequences in the mitochondrial genome, can be constructed from restriction mapping data. However, recombinations between the various repeated sequences (see Fig. 5–8) give rise to a variety of subgenomic circles, as illustrated in the figure. The various repeated sequences are indicated by the open boxes and labeled on the 570 kb master circle with small numbers corresponding to their sizes. Recombinations involving a particular class of repeats are indicated by the arrows connecting the different circular forms, with the recombining repeats indicated by the small numbers next to the arrows. From Lonsdale, and others, *Nucleic Acids Res.* 12:9249, 1984.

in "somatic hybrid" cells produced by protoplast fusion techniques. Simply maintaining cells in tissue culture can lead to variations in the restriction pattern of their mitochondrial DNA. It is difficult from what is known to determine whether these changes reflect recombination events or simply differential replication of different preexisting variant molecules. However, certain somatic hybrid cells have been shown to contain mitochondrial DNA restriction fragments which are not present in either parental cell type but which could be produced by recombination between the parental genomes. And clones have been obtained which contain marker DNA segments from different parents. As yet it is not known whether intergenomic recombination events occur by the same mechanism as the intragenomic recombinations that produce the diverse array of molecules in a single genome, but it seems reasonable to suppose that they do.

**Mitochondrial plasmids.**   In addition to a variety of circular mito-chondrial "chromosomes," mitochondria from a number of plants contain episomal or plasmid-like molecules. These are generally small circular, or small linear, double-strand DNAs. They are usually detected as strong, discrete bands in electrophoretic separations of untreated (intact) mito-chondrial DNA. In well-studied cases, it has been shown that the episomal DNAs do not fit into the restriction map of the main genome. Plasmid-like DNAs have been characterized in a number of plants, including sugar beet, sorghum, and some species of *Brassica*. Normal maize mitochondria carry a linear episome 2.3 kb in length and a circular DNA of about 1.9 kb. This is in addition to the various sub-genomic circles generated by recombination in the main genome. The significance (if any) of the plas-mid-like DNAs to the plant is unknown, and the plasmid-like DNAs can be lost without obvious effects on appearance or viability.

The best studied examples of mitochondrial episomes are found in certain male sterile lines of maize. They have been studied, at least in part, in the hope that they may be involved in producing the male sterility trait. Cytoplasmic male sterility (CMS) appears to be associated with alter-ations in the mitochondrial DNA. Several types of CMS maize cytoplasm can be distinguished by their responses to nuclear "restorer" genes, which restore fertility to some types but not others, and by their mitochondrial polypeptides and the restriction profiles of their mitochondrial DNA. A major CMS cytoplasm is the "S" type, which is characterized by two prom-inant episomal DNAs called S-1 and S-2. In length, S-1 is 6.4 kb and S-2 is 5.4 kb. Both are linear, double-strand DNAs, and both have similar termi-nal inverted repeats 208 bp long. These repeats are covalently linked to protein. By analogy to adenovirus and certain bacteriophage systems, it is thought that the terminal protein complexes may be involved in initiating DNA replication.

The episomal DNAs, S-1 and S-2, are not detectable as free episomes in the mitochondria from CMS-S plants that have reverted to fertility, but sequences homologous to S-1 and S-2 can be found in high molecular weight mitochondrial DNA in normal, in CMS-S, and in fertile revertant cytoplasms. Their arrangement with respect to adjacent ge-nomic sequences differs between CMS-S and revertants, however, and it is thought that this rearrangement may be involved somehow in the process of reversion to fertility. C. Schardl, in collaboration with D. Lonsdale at the Plant Breeding Institute in Cambridge, England and with D. Pring and K. Rose at the University of Florida, analyzed cosmid clones contain-ing sequences homologous to S-1 and S-2 from CMS and revertant plants. Data from restriction analysis of many such cosmids was consistent with a model in which recombination between the terminal inverted repeats of the S-1 and S-2 episomes and homologous sequences in the mitochondrial chromosomes would generate linear mitochondrial DNA molecules containing S-1 or S-2 at their ends. The relationship of this linearization of the otherwise circular mitochondrial genome to the CMS phenotype is still unclear, although the fact that the same arrangements are seen in the DNA of both CMS plants and plants in which the CMS phenotype has been "corrected" by nuclear restorer genes indicates that linearization per se is not sufficient to cause male sterility.

**Chloroplast sequences in mitochondrial DNA.** The observation that chloroplast DNA sequences are contained in the mitochondrial genome provided another surprise for plant molecular biologists. The initial observations came from Cambridge, England, where D. Stern and D. Lonsdale of the Plant Breeding Institute reported that mitochondrial DNA from maize contained a 12 kb sequence from the maize chloroplast genome. When labeled chloroplast DNA was reacted with restriction digests of mitochondrial DNA, this sequence (plus others described subsequently) hybridized preferentially. Further investigation showed that the preferentially hybridizing sequence could be cloned on a cosmid (a plasmid packaged in a *lambda*-virus coat) that contained mitochondrial DNA on either side of the chloroplast segment. The 12 kb sequence turned out to contain a portion of the chloroplast inverted repeat with genes for several tRNAs and the 16S ribosomal RNA. By restriction mapping, the mitochondrial version of the 12 kb sequence appeared virtually identical to its presumed progenitor sequence in the chloroplast, the only differences occurring where the ends of the inserted segment joined mitochondrial DNA sequences.

Two other segments of chloroplast DNA have been characterized in the maize mitochondrial genome. The first of these segments includes the 3' end of the chloroplast 23S ribosomal RNA gene, the genes for 4.5 and 5S ribosomal RNAs, and two tRNA genes. The other segment contains the rbcL gene and its flanking sequences on both the 3' and 5' ends. The gene is functional in an *E. coli* transcription/translation system, and its protein product can be precipitated with antibodies to RuBP carboxylase. However the mitochondrial gene produces a truncated polypeptide of 21,000 daltons instead of the 54,000 dalton protein synthesized by the chloroplast gene. Whether this gene or any other chloroplast gene actually functions in the mitochondrion is unknown. However, the genetic code is slightly different in mitochondria (see subsequent discussion), and there are also likely to be important differences in transcriptional and translational control signals between the two organelles. In view of these differences it seems unlikely that chloroplast genes could be functional in the mitochondrial environment.

The presence in mitochondrial DNA of sequences that hybridize to chloroplast DNA has since been shown to be a widespread phenomenon, which is not restricted to maize. In collaboration with J. Palmer, who had been studying evolutionary relationships among the chloroplast DNAs of a wide variety of plants, D. Stern could show homologies between cloned segments of chloroplast DNA and mitochondrial DNA fragments from several species, including pea, mung bean, spinach, and four different species of cucurbits. Adding up the segments of chloroplast DNA that seemed to be represented in the mitochondrial DNA of one or more of these plants gave the impression that almost the entire chloroplast genome might be subject to random transfer. In addition, different degrees of homology were seen, suggesting that transfer events have occurred at different times during evolution. Although some of the homologies observed in these survey experiments might be the result of cross hybridization between chloroplast and mitochondrial genes of similar function, this is not always the case, and it appears that there are too many cross-reacting fragments to be easily accounted for by this hypothesis. The overall picture is most consistent with a series of events in which random sequences from the chloroplast genome appear at random positions in the mitochondrial genome. These events would be frequent in an evolutionary sense.

There is no direct evidence concerning the mechanism by which DNA transfer occurs between organelles. However, it is easier to reconcile the present indirect evidence with an essentially random process than with directed transfer by some kind of vector. The fusion of organelles or the uptake by one organelle of DNA released by lysis of another organelle might occur frequently enough to explain the observations. This view is also consistent with the recent discovery of mitochondrial DNA sequences in the nuclear genomes of several animal and fungal systems. And it is consistent with the demonstration that chloroplast DNA fragments can be found in the nuclear DNA of higher plants. Sequences transferred between organelles in this way have been called "promiscuous DNA," a term designed to highlight the random nature of the process.

Transfers from mitochondrion to chloroplast have not yet been demonstrated and may not occur at significant frequency. The chloroplast genome simply may not tolerate random insertions of foreign DNA. As noted previously, although rearrangements have occurred in a few chloroplast genomes, chloroplast DNA generally shows a high degree of conservation in both size and sequence arrangement. Those insertion and deletion events which do occur seem mostly to involve rather small segments of DNA which would not be easy to detect by hybridization techniques. In contrast, the large and highly variable mitochondrial and nuclear genomes probably contain many regions in which relatively large pieces of foreign DNA can be inserted with minimal effect.

**Gene content, structure, and expression.**   Mitochondria of higher plants contain substantially the same set of genes as those of other organisms. These include at least one ribosomal protein gene and the ribosomal and transfer RNAs required for the mitochondrial translation system. Mitochondria of higher plants also contain a number of proteins involved in the electron transport and ATPase complexes of the inner mitochondrial membrane. Table 5–3 gives a listing. The list is reasonably complete for yeast and animal mitochondria, for which extensive information is available (including the complete DNA sequences of several animal mitochondrial genomes). In plants, with their very large mitochondrial genomes, it remains possible that additional genes will be discovered. However it seems unlikely that the number of additional genes will be large. When the products of mitochondrial protein synthesis *in vitro* are separated by one- or two-dimensional PAGE, about thirty polypeptide spots can be seen. Some of the spots may be artifacts of the *in vitro* reaction, so thirty spots is probably a maximum estimate. Thus, even though the number of genes listed in Table 5–3 is small, it is likely to represent a reasonably large fraction of all plant mitochondrial genes.

A map showing the positions of several known mitochondrial genes on the 570 kb "master circle" of the maize chloroplast genome is shown in

**TABLE 5-3 Genes identified in plant, yeast, or animal mitochondria.**

|  | Plant | Yeast | Animal |
|---|---|---|---|
| Ribosomal Genes |  |  |  |
| Large subunit | + | + | + |
| Small subunit | + | + | + |
| 5S rRNA | + | − | − |
| Transfer RNAs | 30 | 25 | 22 |
| Ribosome-associated protein | + (1) | + | − |
| Cytochrome c oxidase complex |  |  |  |
| Subunit I (COI) | + | + | + |
| Subunit II (COII) | + | + | + |
| Subunit III (COIII) | ? | + | + |
| Cytochrome *c* reductase |  |  |  |
| Apocytochrome *b* (COB) | + | + | + |
| ATPase complex |  |  |  |
| Subunit F1 alpha | + | − | − |
| Subunit 6 | ? | + | + |
| Subunit 8 | ? | + | + |
| Subunit 9 | + | + | − |
| NADH Dehydrogenase Complex I |  |  |  |
| ND-1 | + (2) | + (?) | + |
| Unassigned Reading Frames | ? | ? | 2 |

Notes:
(1) The "ribosomal protein" has not been definitively identified as such in plant mitochondria, but has been shown to copurify with mitochondrial ribosomes.

(2) The ND-1 gene in plant mitochondria has been tentatively identified on the basis of limited sequence homology with the mouse ND-1 gene.

Figure 5–10. Several differences from the chloroplast genome are immediately apparent, both in the number and in the location of known genes. For example, the genes for mitochondrial ribosomal RNA are arranged in a pattern similar to that in yeast, with the 26S rRNA gene separated by a large distance from the genes for the 18S and 5S rRNAs. These genes are quite close together (separated only by a tRNA gene) in animal mitochondria, and they are part of a single operon in chloroplasts and *E. coli*.

Plant mitochondrial genes also differ from corresponding genes in fungi. An example is found in the presence or absence of introns. The genes for cytochrome *c* oxidase subunit I (COI) and apocytochrome *b* (COB) in fungi contain multiple introns. These have been extensively studied by laboratories interested in the mechanism of RNA splicing since sequences within these introns appear to encode their own splicing enzymes (or maturases). In maize neither COI or COB contains an intron. On the other hand, the maize cytochrome *c* oxidase subunit II gene (COII) contains an intron which is missing from the COII genes of fungi and animals. Wheat mitochondrial COII contains a similar intron, whereas the COII gene from *Oenothera* does not. In maize this intron seems to be spliced out to make a stable, possibly circular, RNA. Little is known about the splicing mechanism except that on the basis of the DNA sequence one may predict that splicing is dependent on RNA secondary structure rather than on specific splicing signals, as in the case of nuclear or chloroplast mRNAs.

The genetic code used by mitochondria often differs from the so-called universal code used by nuclear and chloroplast genes. For example, yeast and animal mitochondria use the triplet TGA (or UGA) instead of

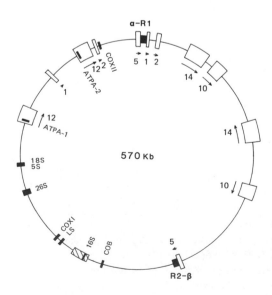

**Figure 5-10.** The positions of various protein-coding genes on the maize mitochondrial genome. S1 and S2 refer to integrated sequences homologous to the S1 and S2 plasmid DNAs found in mitochondria of plants with "S-type" cytoplasmic male sterility. Urf1 (unidentified reading frame 1) is homologous to a gene identified as NADH-ubiquinone oxidoreductase in mammalian mitochondria. LSU and 16S refer to chloroplast sequences in maize mitochondrial DNA (see text). Figure supplied by Dr. D.M. Lonsdale.

the normal TGG to code for tryptophan. In higher plants, mitochondria also appear to use CGG to code for tryptophan. TGA, which encodes tryptophan in the mitochondria of other species, appears to be used as a stop codon in plant mitochondrial genes. In plant mitochondrial genes, there seems to be a strong bias toward the use of codons ending in T, whereas in yeast the bias is in favor of those ending in A or T, and, in animals, those ending in A or C. These differences, especially those involving novel termination codons, would make it difficult to express nuclear or chloroplast genes in the mitochondrial environment.

As in the case of chloroplast genes, mitochondrial genes often produce a complex set of transcripts. An extreme case occurs in animal mitochondria, where a single long transcript is produced. Processing then occurs at the ends of the tRNAs, which are inserted like punctuation marks at the ends of structural genes. Polyadenylation is required to produce functional termination signals. Polyadenylation does not occur in either yeast or plant mitochondria, and transcripts (although often much larger than an individual gene) do not include the entire genome. As yet relatively little is known about transcription or its control in higher plant mitochondria. Internal fragments of all three of the genes mentioned above (COI, COII, and COB) hybridize to a complex pattern of large fragments on Northern blots, suggesting multiple initiation, termination, and/or processing sites. Intron splicing adds to the complexity, especially when, as in the case of the COII gene, the excised intron sequence is sufficiently stable to be detected in hybridization experiments. It is premature to attempt to analyze these complex patterns in much detail, but we can expect to see major advances in this area as more genes are identified, mapped, sequenced, and used to analyze the synthesis and processing of mRNA.

# SIX

# Manipulating the Genome

Modern techniques for isolating and analyzing eukaryotic genes have raised hopes (and fears) that these genes can be used to transform plant and animal cells. Transformation is the insertion of foreign genes into a cell, and it has been observed in bacterial systems for many years. Transformation of plant cells might give us crops that have previously unknown combinations of traits—corn that synthesizes ammonia from atmospheric nitrogen is a well-used example. Other possibilities of transformation include changes in the amino-acid composition of seed storage proteins, increases in the efficiency of photosynthesis, and new resistance to viruses, bacteria, fungi, and insect predators. For these reasons, many commercial firms are sponsoring research aimed at facilitating the transformation of plant cells.

Transformation and related processes for changing a cell's genetic complement or the arrangement of the genes in its chromosomes also provide a new experimental strategy for studying development. The strategy will be particularly useful for determining the function of genes that control development. For example, is there a single gene postulated to produce leaves with Kranz anatomy, the tissue organization seen in C4 plants? Add this gene to a C3 plant, which normally does not have such organization, and observe the result. Is there a DNA sequence thought to promote transcription in the light? Connect this DNA sequence to a foreign gene and insert the complex into a cell known to respond to light. Then observe whether or not the foreign gene's transcripts appear with those of other light-controlled genes. Do you believe that a particular gene inhibits the expression of nearby genes in root tissue? Add a copy of this gene at another place in the genome, near a gene that is normally expressed (but not obligatory) in roots, and see whether or not the expression of that gene is turned off. In the past experimental changes in the genome were confined to random mutations and were generally destructive, involving the loss of a gene's function. Now changes may be constructive, providing a gain in function, and they need not be random in structure or in position within the genome.

The difficult (and interesting) techniques needed for transforming plant cells involve, for the most part, the insertion of DNA into the cell and the incorporation of the DNA into a replicating chromosome. Techniques for the preparation and analysis of particular gene sequences have already been described in previous chapters. While such preparations of pure DNA can be injected into cells with micropipettes or inserted using other methods such as electroporation (see the technical insert at the end of this chapter), large scale, efficient incorporation of new DNA into chromosomes is most often performed using a "vector" DNA, analogous to the plasmid and viral DNAs used to transform many bacteria. There are three main vectors that are now being investigated. Two of these, the Ti plasmid of *Agrobacterium tumefaciens* and cauliflower mosaic virus (CaMV), are natural plant pathogens. The third, a transposon, is a type of DNA that is present naturally in some plants and known to insert at various places within chromosomes. It is possible to use these to carry foreign DNA into plant cells.

## A. TRANSPOSONS

Transposons are special sequences of DNA that are found in the genomes of many species, including *E. coli,* fruit flies, maize, and snapdragons. Their most spectacular characteristic is the ability to move from place to place within the genome, being excised from one site on a chromosome and inserted in another site, generally at unpredictable times. Excision and insertion can affect the expression of genes near the places to which and from which they move. In general, the insertion of a transposon into the coding sequence of a gene will prevent the expression of that gene by interfering (at least) with translation. Insertion into a site outside the coding sequence but within a regulatory region (for example, a promoter) may interfere with transcription. The excision of a transposon from its site within or adjacent to a gene may allow renewed expression of the gene, but also it may not (see following discussion).

The transposons of bacteria have been studied extensively. The smallest and simplest ones contain two genes, flanked by short, identical or nearly identical base sequences oriented in opposite directions to each other (inverted terminal repeats). The two genes code for enzymes that cooperate to catalyze the insertion and excision processes. More complex transposons have extra DNA, including additional genes, between the terminal repeats, or they have DNA flanked by two smaller and simpler, but complete, transposons.

Transposons were discovered in maize through the work of Ralph Emerson on mutant reversion and the work of Barbara McClintock on chromosome breakage. McClintock's story is particularly interesting. In the 1940s and 1950s, she recognized and reported the movable nature of the transposons, as well as some of their complex regulatory properties, but her reports were considered more curiosities in genetics than phenomena of basic significance. The importance of McClintock's "jumping genes" was generally recognized only after similar elements were found and characterized in bacteria. Now, there is speculation that transposons are responsible for much of the repeated DNA in eukaryotic genomes and for much of genome evolution. Since the variation in genome size and repetitive structure is so much greater among plant species than animals, it may be that there are more, or more active, transposons in plants.

A pictorial example of transposition is found in maize seeds that are homozygous for a transposon mutation in one of the eight genes required for anthocyanin synthesis. Such seeds are yellow unless one of the copies of the transposon-inactivated gene reverts, producing anthocyanins which make the seeds turn red or purple. If the reversion occurs during the seed's development, then only those cells produced after the reversion are colored, leaving a red spot on the yellow seed coat. The size and number of the spots show the timing and frequency of transposition out of the anthocyanin gene. Frequent transposition gives more spots; early transposition gives larger ones (Fig. 6–1).

Another example of a transposon in maize is the Robertson's mutator gene (Mu). A strain of maize that carries Mu somewhere in its genome will, with a frequency 50 times higher than average, produce progeny that contain mutations in any of several other genes. Michael Freeling at the University of California at Berkeley has used a Mu-containing strain to produce plants deficient in alcohol dehydrogenase. These plants are deficient because a copy of Mu has inserted itself in the first intron of the Adh-1 gene. Such plants will also produce, with higher than average frequency, revertant, wild-type progeny as the Mu leaves the Adh locus (Fig. 6–2).

McClintock worked with transposons that were more complex than Mu. One type (called Ds) could move through the genome and cause chromosome breaks at the points where it inserted, but this happened only in the presence of a related type of transposon (called Ac). Ac itself could move through the genome with or without Ds being present. Thus,

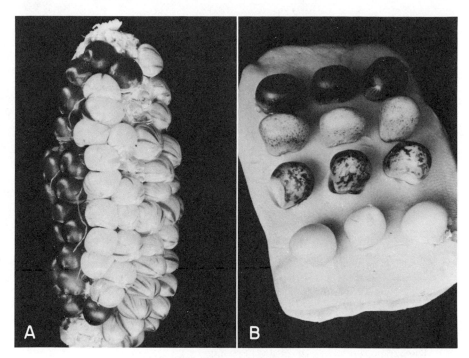

**Figure 6-1.**   Effects of transposons on maize seed color. A. Reversion of the inactivation of pericarp color gene Mp by the transposition of Ac out of the gene. B. Breakage of a chromosome containing Ds (in the presence of Ac). The breakage results in the loss of the color-inhibitor allele, $C_I$. In both A and B, the number and size of the colored regions reflects the frequency and timing of the events. Photographs courtesy of Dr. John I. Yoder.

Ac was an "autonomous" and Ds was a "non-autonomous" transposon. There is evidence that Ds is a defective form of Ac (discussion follows).

It has been possible to isolate and determine the structures of the Mu, Ac, and Ds transposons. Investigators first constructed a cDNA probe to a known maize gene; then they located (on a Southern blot) the corresponding sequence in a transposon-mutated strain of maize. The base sequence of the mutant gene was compared, base by base, to that of the wild-type gene. Each mutant sequence contained a sequence of bases inserted into or near the wild-type gene (Fig. 6–3). These insertions differed according to the type of transposon, but they were similar in having inverted repeats at their ends. For instance, the Mu insertion in Adh-1 studied by Freeling and Taylor and their colleagues had a sequence of about 235 base-pairs at one end, which was repeated (with some differences) at the other end. The Ac insertion studied in the maize waxy locus (wx-m7) had a region of about 150 base pairs repeated at each end,

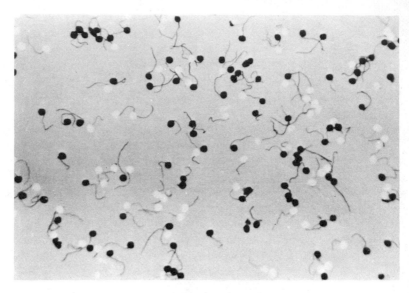

**Figure 6-2.** Detection of Adh+ and Adh− maize pollen. Pollen shed from a heterozygous plant were germinated, frozen, and stained with NBT, which turns blue when reduced by alcohol dehydrogenase. Revertants of ADH-strains can be detected as blue-staining pollen from an Adh−/Adh− plant. Photograph courtesy of Dr. Michael Freeling.

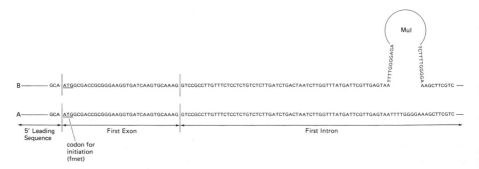

**Figure 6-3.** The base sequence of a part of the maize Adh-1 gene, shown (A) without and (B) with the Mu1 insertion. See J.L. Bennetzen, J. Swanson, W.C. Taylor, and M. Freeling, *Proc. Natl. Acad. Sci.* USA 81:4125, 1984.

though again the repeat was imperfect. At the very ends of Ac, there were perfect repeats of eleven base pairs (Fig. 6–4). Ds transposons had the same inverted repeats as Ac. It is likely that the inverted repeats form the essential parts of the transposons that allow them to move in and out of the genome.

The interior part of Ac (wx-m7), between the inverted repeats, contains three "open reading frames" (ORF). An ORF is a sequence starting

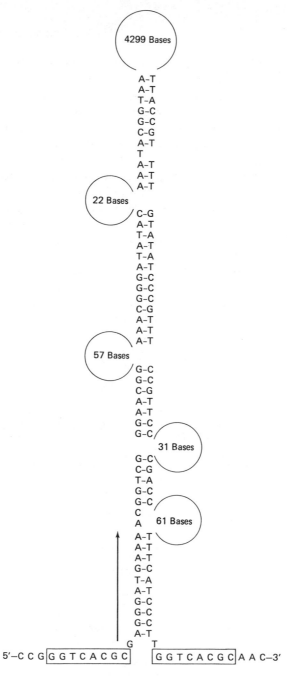

**Figure 6-4.** Structure of Ac in the maize wx-m7 mutant. The vertical segment represents the terminal inverted repeat of the transposon; the top circle represents the transposon's central sequence, containing three open reading frames. Most Ds elements contain a complete inverted repeat but lack some part of the central sequence. At least one Ds element (Ds1) has only the terminal 13-19 bases of Ac (plus a central sequence unrelated to Ac). The horizontal segment at the bottom is host DNA.

with ATG, continuing through an integral number of base pair triplets coding for amino acids, and ending with a termination codon. Each ORF can be considered a potential gene, although of course without further evidence we cannot know if it is really expressed as a protein. Ds insertions differ from Ac in their lack of these ORFs. Some forms of Ds lack a small part of Ac, some a large part. Some have duplicated segments of Ac and are larger than Ac, even though they are incomplete. Some have quite foreign DNA between their Ac-like repeats. Remembering that Ds cannot transpose without Ac present and active, we can suggest that the genes (ORFs) in Ac encode for enzymes that catalyze the insertion and removal of DNA that has the particular Ac-Ds-like inverted repeats.

In comparing transposon-mutated genes with wild-type genes, one notes that there are direct repeats at the ends of the inserted sequences. These are sequences of the wild-type gene, although in the wild-type gene there is only one copy. The duplication results from the insertion process—at least we assume so from the studies of transposition in bacteria. As shown in Figure 6–5, it is suggested that when the insertion is made

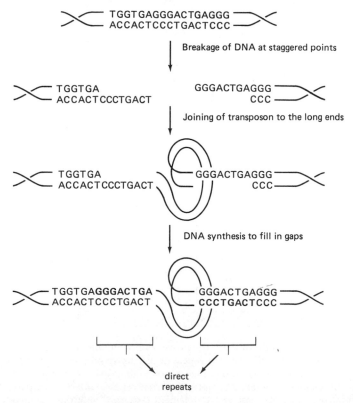

**Figure 6-5.** Formation of direct repeats in DNA following the insertion of a transposon.

there is a staggered cut in the two chains of the target DNA, such as that made by many restriction enzymes. The transposon DNA is attached to the long chains, and the gaps are filled in by a repair DNA polymerase. When transposons move away they usually leave the direct repeat in place.

The importance of transposons for genetic manipulation lies in their ability to integrate into DNA. It is possible that a transposon could be modified by adding a desired gene between the inverted repeats. If this DNA could be placed in a cell, perhaps by electroporation (see technical insert on electroporation at the end of Chapter 6, Section D), it might efficiently insert itself into one or many sites in the genome.

## B. CAULIFLOWER MOSAIC VIRUS

There are many types of plant viruses, but most of those studied have a genome made of RNA. Cauliflower mosaic virus (CaMV) represents a type of virus that contains DNA (Fig. 6–6). This means that it can be studied and modified by the usual techniques of molecular biology. And perhaps it can be used as a vector to introduce different genes into plant cells.

CaMV infects cauliflower and a few other members of the Cruciferae. In nature it is spread from plant to plant by aphids, which pick up virus particles as they feed on infected plants and reinject the virus into new leaves. In the laboratory, virus particles (or the DNA isolated from them) can infect a plant if they are rubbed on the surface of a leaf in the presence of a small amount of abrasive compound. Upon infection, the virus particles reproduce in the cells. Their DNA genomes serve as templates for the synthesis of new DNA and for the transcription of messenger RNA, which encodes for coat protein and other proteins that might be needed for the maturation and spread of the virus.

In an infected plant, virus particles move through the phloem into the expanding leaves where they disturb leaf development. These leaves have a mosaic or mottled appearance, with vein clearing and dark green islands in the midst of more normally colored regions. Differences in the expansion rates of the different leaf regions lead to a crinkled and stunted leaf appearance. How the virus produces the changes in leaf morphology is itself an interesting problem in development, but since the same symptoms are caused by many other viruses it is unlikely that unraveling the specific characteristics of CaMV will help much in solving this problem.

Within the cell, infection by CaMV is associated with the appearance of inclusion bodies, and proliferation of the virus probably occurs within these bodies.

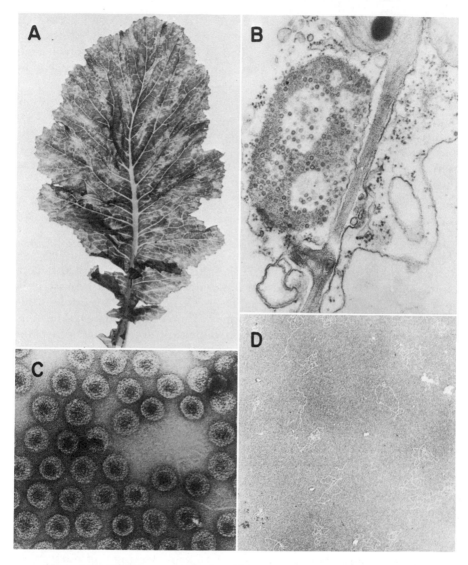

**Figure 6-6.** Cauliflower mosaic virus (CaMV). A. Infected leaf. B. Virus particles in an inclusion body. C. Purified virus particles. D. Viral DNA circles. Photographs by Dr. Robert J. Shepherd.

The CaMV particle is spherical and about 0.05 micrometers in diameter. Its core of DNA is surrounded by a coat made of two kinds of protein. The DNA itself is a double-strand circular molecule containing 8024 base pairs (Fig. 6–7). Both strands of the DNA have breaks and overlaps that seem to be consistent and important parts of their structure, but as yet no one knows why. The whole base sequence has been de-

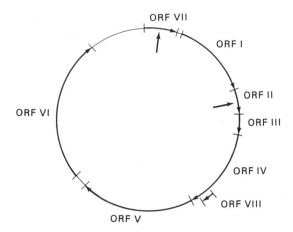

**Figure 6-7.** Schematic diagram of CaMV DNA. Roman numerals next to heavy lines indicate genes (ORF, open reading frame); arrows show two sites that can be modified without destroying the infectivity of the viral genome. Redrawn from Howell, *Annu. Rev. of Plant Physiology* 33:609, 1982.

termined, and from the sequence one can identify eight open reading frames (ORFs). So far, the genetic evidence suggests that ORF VI codes for the matrix protein of the inclusion body. ORF IV codes for a coat protein. ORFs II and VII seem not to be necessary for a successful infection.

Inside infected cells are found two major RNA transcripts. One covers the sequence of ORF VI. Another and larger one contains the base sequence of the whole genome. (The RNA polymerase goes all the way around the template strand and then continues for a short distance, repeating the first transcribed segment.) It is likely that the larger RNA acts as a polygenic messenger RNA. This suggests that any gene that is inserted into the genome will be transcribed. Genes that are inserted in the correct direction may well be translated; however, it is possible that the genes contain some special information for initiating translation besides the base triplet AUG. This special information may be something like the Shine-Delgarno sequence, even though this sequence is not found in most eukaryotic mRNAs.

From our knowledge of CaMV's structure and its infection process, we can guess that this virus might help researchers introduce genes into plants in a way that would encourage their spread and their expression. A gene inserted into the CaMV genome could be replicated, packaged, and transmitted through the phloem. It could then be transcribed and translated along with the other CaMV genes so long as it did not increase the physical size of the viral genome beyond the size which could be packaged by the viral coat.

There are certain limitations on the use of CaMV as a vector. First, there is nothing to suggest that a gene in CaMV is likely to be integrated into the plant's genome. Therefore, it is unlikely to be transmitted from generation to generation through seed. Second, it is possible that the symptoms of viral infection may hinder the expression—or the study of the expression—of the genes of interest. Third, CaMV has a restricted host range, which means that it cannot be used with many experimentally or commercially useful plants. With these limitations in mind, CaMV can be a useful experimental tool.

## C. AGROBACTERIUM TUMEFACIENS

*A. tumefaciens* is a soil-borne bacterium. Like *Rhizobium,* the nitrogen-fixing bacterial genus, *Agrobacterium* has developed a way of living in and deriving nourishment from plant tissues. However, unlike *Rhizobium, Agrobacterium* is a parasite and provides no benefit to the plant that it colonizes. Instead, it causes crown gall disease.

*Agrobacterium* can infect many dicot—but no monocot—species. In the normal disease process the bacterium enters the plant at the site of a wound. The infection often occurs at the crown (the junction of the root and shoot), but it may involve an upper part of the stem, the petiole, or the leaf. The bacterium attaches to the wall of a cell and "transforms" the cell. In this case, the word "transforms" was used before it was known to involve DNA transfer, referring only to the observation that infected cells acquire new properties. Infected cells proliferate at the site of infection, forming a tumor or "gall." The tumor can be fatal to young plants if it grows big enough to crush the vascular system, although often the tumor is merely debilitating and unsightly. Transformed cells also have the ability to grow in culture on a medium that lacks the hormones auxin and cytokinin. Finally, transformed cells produce opines, such as octopine and nopaline. These are chemicals that are formed from two amino acids. Argenine and alanine form octopine, and argenine and glutamine produce nopaline. Octopine and nopaline are not found in normal plant cells.

There are two more or less distinct types of crown galls (Fig 6–8). The first type grows as a relatively unorganized callus on artificial medium and on host plants. The cells of this type produce octopine. The second type grows as a callus containing green islands. The islands show a variable amount of organization, including the production of multiple shoots. The cells of this type produce nopaline. The two types are associated with different strains of *A. tumefaciens.* There is also a related disease ("hairy root disease") in which infected tissues proliferate in root tissue and produce an opine. This disease is associated with the bacterium *A. rhizogenes.*

From the variety of tumors formed, it is clear that the pathogenic bacteria influence the development of infected tissue. Similar types of de-

**Figure 6-8.** Crown-gall callus tissues. Photograph courtesy of Dr. M.P. Gordon. Reprinted from *The Biochemistry of Plants, A Comprehensive Treatise,* P.K. Stumpf, E.E. Conn, eds., Vol. 6, "Proteins and Nucleic Acids," A. Marcus, ed., Academic Press, 1981.

velopment are commonly observed in normal tissues cultured on artificial media (Chapter 1, Section B). The different developmental patterns (unorganized, shoot-forming, and root-forming) are seen in media that contain different combinations of plant hormones. A medium with a balanced mixture of auxin and cytokinin tends to promote the growth of unorganized callus, whereas a medium with high cytokinin and low auxin stimulates shoot induction. On the other hand, a medium with high auxin and low cytokinin promotes root development. By comparing tumor types with the developmental pattern of normal tissues on these media, we can hypothesize that infected tissues have been induced to synthesize auxins and cytokinins. Moreover, we can guess that unorganized tissue forms a balanced mixture of the two hormones; also, shoot-forming tumors form less auxin, and root-forming tissues make more auxin (or less cytokinin). In general, these hypotheses have been confirmed by chemical analysis of the hormone contents of tumor tissues.

The formation of opines explains the ecological significance of tumor formation. Each strain of *Agrobacterium* synthesizes enzymes (permease, dehydrogenase) that allow it to metabolize the specific type of opine formed by the tumor it induces. Thus, by stimulating the plant to form opines, the bacteria insure themselves a supply of nutrients specifically designed for them. The growth of the infected tissue is important because it increases the amount of opine-forming tissue. It is possible, too, that the physical characteristics of the tumors—parenchymal cells with extensive intercellular spaces—provide a good habitat for the bacteria.

Early studies into the mechanism of the "transformation" process demonstrated that tumorous tissue remained transformed, even in the absence of the infecting bacteria. (The bacteria could be removed with antibiotics like penicillin.) Armin Braun at Rockefeller University reasoned that the bacteria had provided a "tumor-inducing principle." The nature of this principle remained obscure until the 1970s, when Jeff Schell and his co-workers at the State University of Ghent in Belgium found that virulent bacteria all contain a large circular DNA plasmid. When strains of bacteria lose the plasmid, they lose the ability to infect plants. When bacteria regain a plasmid, they regain virulence. Further experiments by Rob Schilperoort's group in Leiden, Holland, demonstrated that many of the genes that control tumor characteristics (morphology, opine synthesis) lie on the plasmid. Finally, Mary-Dell Chilton, Eugene Nester, and Milton Gordon and their associates at the University of Washington in Seattle showed that a segment of the plasmid, the T ("transferred") DNA, integrates into the genome of transformed cells. The plasmid ("Ti plasmid") is thus identified as the tumor-inducing principle.

The initial experiments that attempted to detect the presence of Ti plasmid DNA were apparent failures. These experiments measured the rate of reassociation of denatured radioactive Ti plasmid DNA in the presence and in the absence of DNA from crown gall tumor cells. In theory, if the crown gall cell DNA contained plasmid DNA sequences (the hypothetical T-DNA), then adding crown gall cell DNA should accelerate the rate of reassociation of those sequences. At first, no acceleration was detected. The problem turned out to be the large size of the Ti plasmid and the small size of the T-DNA. Only a small fraction of the radioactive Ti plasmid reassociated faster in the presence of crown gall DNA, and this fraction was too small to detect. When the test was repeated using *Sma* I restriction fragments of the Ti plasmid, two fragments were found to reassociate rapidly in the presence of crown gall cell DNA. Later experiments, using radioactive Ti plasmid fragments as probes on Southern blots of crown gall DNA, confirmed the conclusion that crown gall cells contained T-DNA sequences from the Ti plasmid.

There are many powerful techniques for analyzing the structure of plasmids and most have been applied to the Ti plasmids. Figure 6–9 diagrams two types of Ti plasmids, one which induces octopine-synthesizing tumors and one which induces nopaline-synthesizing tumors. The two have very similar physical and genetic structures. They are large circles, measuring 140–235 kilobase pairs. The two plasmids have genes for transfer of the plasmid between bacteria, for opine and arginine catabolism, for assuring that there will not be two types of plasmids in the same cell, and for excluding a certain bacteriophage. They also have genes outside the T-DNA that are necessary for virulence. It is not known exactly what these control, but it might be attachment of bacteria to plant cell walls, transfer of the plasmid into a plant cell, or integration of the T-DNA into the plant genome. Finally, the two plasmids have the T-DNA itself.

As described previously, the T-DNA is identified by isolating the DNA from bacteria-free cells and locating the sequences that will hybridize to Ti-plasmid DNA. In every case, the T-DNA represents a single,

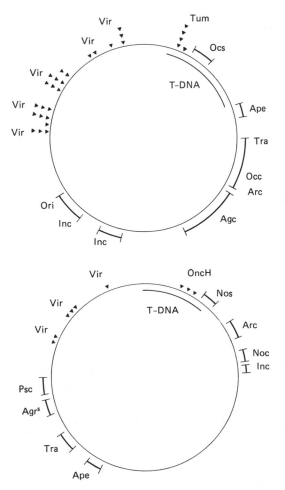

**Figure 6-9.** Structure of Ti plasmids. Genes are abbreviated as follows: Agc, agropine catabolism; Agr-s, agrocin sensitivity; Ape, bacteriophage API exclusion; Arc, argenine catabolism; Inc, incompatibility; Noc, nopaline catabolism; Nos, nopaline synthase; Occ, octopine catabolism; Ocs, octopine synthase; OncH, host range; Ori, origin of replication; Psc, agrocinopine production; Tra, transfer; Tum (tmr, tms, tml), tumor morphology; vir, virulence. Reproduced with permission from the *Annual Review of Microbiology*, Vol. 35, © 1981 by Annual Reviews, Inc.

contiguous region of the plasmid. This region is often moved into the plant genome as a single piece, but in some tumors the T-DNA is split in two (into T-L and T-R) and integrated into the plant DNA in two separate places.

The T-DNA includes several genes (Fig. 6–10). Some of these have been identified only by the fact that they produce nRNA transcripts that

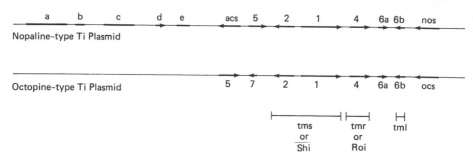

**Figure 6-10.** A. Structure of T-DNA in plasmids. Numbered arrows represent mRNA-producing genes. Arrowheads show direction of transcription, when known. Redrawn from Caplan and others, *Science* 222:815. Copyright 1983 by the AAAS.

can be isolated from the tumor cells; the functions of the proteins coded by these transcripts are generally not known. But the functions of some of the genes have been determined by analyzing the phenotypes of transposon-insertion mutants. Two genes, ocs and nos, direct the synthesis of octopine or nopaline, respectively. Another gene, found in nopaline synthesizing plasmids, but not in octopine-synthesizing plasmids, directs the synthesis of another opine, agrocinopine.

There are at least four T-DNA genes that influence tumor morphology. The first two, called tms or Shi genes, apparently inhibit shoot formation since mutants of these genes induce tumors that form shoots. The third gene, called tmr or Roi, apparently inhibits root formation. Mutants in a fourth tumor-morphology gene, called tml, produce unusually large tumors. Studies of these mutants confirm the hypothesis that the balance of hormones determines growth and development in these tumors. Tumors from mutants of the tms genes have a ten-fold higher ratio of cytokinin to auxin than wild type tumors. These tumors need auxin for optimum growth in culture. Tumors from tmr mutants require cytokinin for optimal growth in culture. The tmr gene codes for the enzyme isopentenyl transferase, the first enzyme in the cytokinin pathway.

From the point of view of someone wanting to insert genes into a plant cell genome, the genes that control integration of the T-DNA into the plant DNA are among the most interesting. Studies with modified Ti plasmids show that the border regions of the T-DNA are necessary for integration (especially the right hand region). There is a 25-base-pair sequence directly repeated at the left and right ends of the T-DNA, and this sequence may be involved in the mechanism of integration. The internal genes of the T-DNA are not necessary for integration, although they are necessary for tumor formation. The internal genes can be replaced with "foreign" genes, which are then integrated under control of the border regions. This provides a mechanism for transforming cells with a wide variety of genes.

## D. EXAMPLES OF TRANSFORMATION

One report of the insertion of a foreign gene into a plant describes a use of CaMV. In this experiment, a gene from *E. coli* was inserted into the CaMV genome in place of a gene that was unnecessary for viral replication. CaMV-infected turnip cells demonstrated the presence and expression of this gene in several ways.

The gene chosen for introduction was the dihydrofolic reductase gene (DHFR) from *E. coli* strain 67. The enzyme coded by this gene is very resistant to the antibiotic methotrexate, whereas most other DHFR enzymes are inhibited by this compound. The investigators, Brisson and his colleagues from the Frederick Miescher Institute in Switzerland, removed most of open reading frame II with restriction enzymes and ligated the DHFR gene in the same place (Fig. 6–11). Brisson took particular care to reduce the number of excess intergenic bases between the DHFR gene and the adjacent CaMV genes (I and III), and this turned out to be important for the reproduction of the virus.

Infected leaves, ones that developed several days after viral DNA was rubbed on older leaves, contained the *E. coli* DHFR. This was demonstrated by a Western blot, showing that the cells had an enzyme protein that reacted with antibody to the *E. coli* enzyme. Uninfected leaves did not have the antibody-reactive protein. The plants infected with virus were also resistant to methotrexate. They could incorporate $^{32}$P into DNA in the presence of this drug and survive several spray treatments. Control plants were unable to synthesize DNA in the presence of the drug, and they died.

This experiment shows the feasibility of inserting and expressing a foreign gene, in fact a prokaryotic gene, in a plant cell. As we might expect, the gene was present only in infected cells, and there was no evidence that it moved into a stable position in the plant genome. In fact, when a larger piece of DNA containing the DHFR gene was inserted into the CaMV DNA, it was unstable even in the viral genome, and it tended to disappear, being undetectable in the progeny virus.

The Ti plasmid system has also been used to effect transformation of plant cells. This has been developed carefully and is so flexible and effective that it shows promise of becoming the routine tool for genetic engineering of plant cells.

The advantage of Ti plasmid as a vector is its ability to integrate into the nuclear genome of its host in the same manner as a transposon but apparently in a more stable fashion. Thus, unlike CaMV-infected plants, Ti-infected plants might be able to maintain foreign genes in a stable form and transmit them sexually to their progeny.

The disadvantage of Ti plasmid as a vector is its tumorigenic property, which alters the auxin-cytokinin balance in host cells and makes it difficult to regenerate fertile plants. This problem can be alleviated by

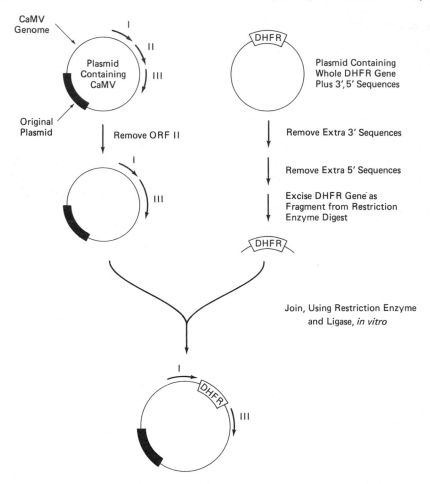

**Figure 6-11.** Strategy for construction of a hybrid CaMV designed to introduce a mutant dehydrofolic acid reductase (DHFR) into plant cells.

constructing a T-DNA that does not contain the tumorigenic genes. Such a T-DNA has been constructed and cloned by at least three research groups. One small plasmid created by Horsch and colleagues at the Monsanto Company contains a T-DNA with a nopaline synthetase gene and the right-hand border sequence of T-DNA that is essential for integration. It also has a segment of DNA that is the same in nopaline-type and octopine-type Ti plasmids; this segment allows the insertion of the new T-DNA into an octopine-type Ti plasmid by recombination. It contains a gene for resistance to streptomycin that aids in selecting the hybrid plasmids that contain the new T-DNA. Finally, it has a gene for neomycin phosphotransferase II. This gene makes the plant cells resistant to the antibiotic kanamycin. The kanamycin resistance trait can be used for selection of transformed plant cells.

In the 1984 study performed by Horsch and his colleagues, an *A. tumefaciens* strain with an octopine-type Ti plasmid was infected with the small plasmid containing the new T-DNA (Fig. 6–12). This plasmid was not stable unless it recombined with the Ti plasmid, so a selection for streptomycin resistance picked out recombinants. The recombinants had Ti plasmids with one left-hand and two right-hand border sequences. When these plasmids infected *Nicotiana plumbaginifolia* cells, at least some of the infections resulted in the transfer of only the new T-DNA genes to the plant genomes. Colonies of such cells could be identified by their resistance to kanamycin and their ability to regenerate plants. Nontransformed cells were not resistant to kanamycin; and cells that received a larger T-DNA sequence, including the tumorigenic genes, could not regenerate plants. Of all the cells that were exposed to the *A. tumefaciens,* about 1% were transformed; of all the transformants, about 10% were able to regenerate.

A special feature of Horsch's study was the demonstration of sexual transmission of the kanamycin-resistance gene. Flowers from regenerated, kanamycin-resistant plants formed seed following self-fertilization. The seeds were planted, and the new seedlings tested for resistance. About three-quarters were resistant to kanamycin, as would be expected if a single, dominant gene had been added to one chromosome of the parent plant's cells. The experiment shows the stability of the genes introduced by Ti-plasmid transformation.

In view of the demonstrated usefulness of the Ti system, we can now look for ways to use it to alter development and test hypotheses about developmental control. The first steps have already been taken: genes that are normally expressed at particular stages of development have been combined with different promoters to see whether or not the promoters affect the patterns of their expression.

In an experiment reported by Murai at the University of Wisconsin and his collaborators from Agrigenetics, Inc., a cloned gene for phaseolin, the storage protein in bean seeds, has been inserted into cultured cells of sunflower (Fig. 6–13). Three different hybrid Ti plasmids were used to infect the sunflower cells. The first, the phaseolin gene, minus its promoter and its first 11 codons, was fused to the promoter and first 88 codons of the Ti plasmid octopine synthetase gene. The octopine synthetase gene is normally expressed in all crown gall tissue. In the second plasmid, the entire phaseolin gene with its own promoter was placed in the middle of the tml gene, with the 5′ terminal end of the phaseolin gene toward the 5′ end of the tml gene. The third plasmid was constructed like the second except that the 5′ end of the phaseolin gene pointed toward the 3′ end of the tml gene. The tml gene is also normally expressed in

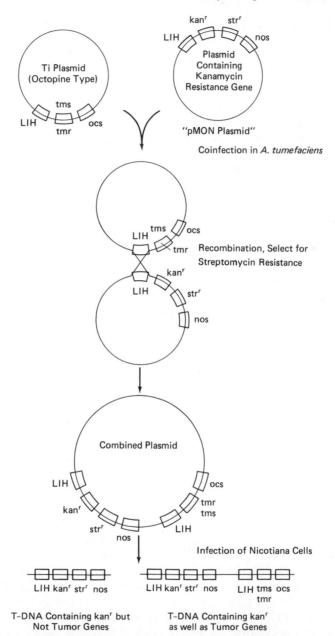

**Figure 6-12.** Strategy for construction of a hybrid T-DNA designed to introduce the gene for kanamycin resistance into *Nicotiana plumbaginifolia*. "LIH" is a region of homology between the two plasmids, allowing recombination. "kan-r" is the gene for kanamycin resistance. "str-r" is the gene for streptomycin resistance. "tmr" and "tms" are tumor-forming genes. "ocs" and "nos" are genes for octopine synthase and nopaline synthase.

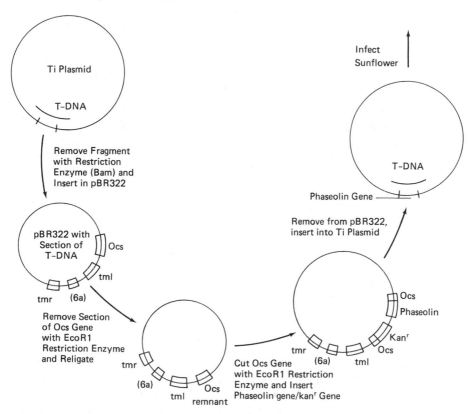

**Figure 6-13.**  Strategy for construction of a hybrid T-DNA designed to introduce a phaseolin gene into sunflower cells.

crown gall tissue, but we might expect that the expression of the phaseolin gene in the second and third plasmids would be controlled by the phaseolin promoter, which is normally turned on only in developing seeds and not in crown gall tissue.

Crown gall tissue that was infected with the first plasmid (octopine synthetase promoter) had sequences among its poly(A)+ RNAs that reacted with phaseolin cDNA and directed the synthesis *in vitro* of protein that reacted with antibody to authentic phaseolin. In this situation, the octopine synthetase promoter (and not the phaseolin gene, its introns, or its 3′ adjacent sequences) apparently controlled transcription. Crown gall tissue that was infected with the other two plasmids (phaseolin promoter) also produced phaseolin poly(A)+ RNA, but in much lower quantities, about one-twentieth as much as with the first plasmid. This observation is consistent with the idea that, in these cases also, transcription was controlled by the promoter. However, the amount of RNA, while low, was much more than would have been found in bean leaves, and the amounts of RNA in galls induced by the second and third plasmids were not always equal. Thus it is possible that sequences outside the gene, as well as species- or tissue-related factors, also influenced the amount of transcription.

Whereas it was possible to detect anti-phaseolin-reactive protein in the crown gall tissue, the amounts were very low. It was found that proteases quickly degraded the phaseolin produced by translation. Breakdown of this gene product is another level of regulation of gene expression, though perhaps not a natural one in bean plants.

The experiments with phaseolin described above were performed with a Ti plasmid containing tumorigenic genes. Therefore it was not possible to regenerate sunflower plants and study the expression of the inserted genes in different tissues. Subsequently, tobacco cells were transformed with "disarmed" Ti plasmid containing a phaseolin gene. Plants were regenerated from these cells and induced to flower. Phaseolin appeared primarily in the protein bodies of the seeds of these plants and not in other tissues. Apparently the phaseolin gene that was transferred contained regulatory elements that assured the correct localization of the gene product at both the tissue and subcellular levels of organization, even in the genetic background of a plant from an entirely different family.

The Ti plasmid transformation system has also been used to investigate light-induced transcription of the small subunit of rubisco. Several different research groups have demonstrated that the small subunit promoter can function in a host cell. In 1984, Broglie, together with colleagues from Rockefeller University and the Monsanto Company, inserted an entire pea small subunit gene into petunia protoplasts. Four or five copies of the pea gene persisted in each cell of the callus that grew from the protoplasts. The pea gene was transcribed in the petunia cells. Several lines of evidence suggested that transcription was controlled by the pea gene's own promoter. For instance, the size of the transcript was the same regardless of the right-handed or left-handed orientation of the pea gene with respect to the other genes of the plasmid vector. Transcripts of the small subunit gene were at least 50 times more prevalent in light-grown than in dark-grown petunia callus cells. Another transcript from the Ti plasmid, synthesized in the same cells but controlled by the nopaline synthetase promoter, was not affected by light, showing that the light effect on the small subunit gene was specific. A different experiment, reported by Herrara-Estrella and colleagues, showed that the promoter from a different pea small subunit gene could be linked to a chloramphenicol acetyltransferase (cat) gene and inserted into tobacco protoplasts through a Ti system. Expression of the cat gene was stimulated 5-fold to 15-fold by light. Apparently, the small subunit genes contain light-regulated promoters which maintain their properties even in foreign cells.

The sequences that regulate small subunit gene expression can be located by modifying the genes used in transformation experiments. The small subunit gene that was used in the petunia cell experiments described previously contained 1,052 base pairs upstream (5') from the origin of transcription. Various segments of the 5' region can be deleted so that shorter lengths are inserted into the host protoplasts (Fig. 6–14). When this experiment was performed, a Northern blot analysis of the RNA that was transcribed from the modified genes showed a pattern in dark and light that confirmed the promoter activity of the 5' sequences. Deletion of 700 bases, leaving 352 bases flanking the gene, cut the level of mRNA found in light-grown callus by more than a factor of five. Deletion of the 22 bases around the TATA box removed promoter activity entirely. From these and other observations shown in Figure 6–14, we can conclude the following: (a) Transcription is promoted by sequences around the TATA box, within thirty-five base pairs of the start of transcription. (b) Such transcription is light-regulated. The actual base se-

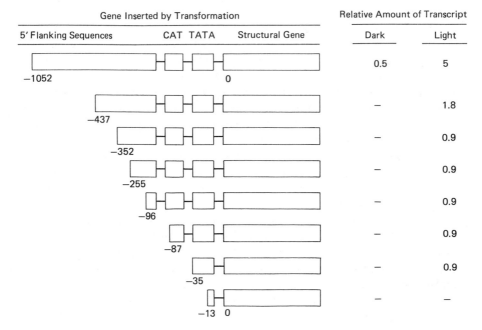

**Figure 6-14.**   Light-regulated transcription of a pea rubisco small subunit gene in petunia callus. A gene clone in pBR325 was modified by deletion of 5' flanking sequences, then recloned and recombined into Ti plasmids using an intermediate vector that infects both *E. coli* and *A. tumefaciens*. The left hand column diagrams the modified segments of the small subunit gene that were present in the Ti vector; it is not drawn to scale. The figures in the right hand column show the relative amount of small subunit gene transcript accumulating in darkness and in light: "-" means that transcript was not detectable.

quences within this region that confer the light response have not yet been reported, but there is a directly repeated sequence that looks interesting. (c) In this gene the CAAT box sequences, though present, are not required for promoter activity. This does not mean, however, that they have no effect in their normal environment. (d) Transcription is enhanced by sequences over 400 base pairs from the 5' end of the gene.

Similar Ti transformation techniques are being used to study the properties of transit peptides that cause proteins to enter chloroplasts. This is done by attaching segments of the small subunit gene to the gene of a protein that would not normally enter the plastids. A fusion of the gene for the small subunit transit peptide (both with and without sequences that code for more of the small subunit) to the gene for neomycin phosphotransferase has been tested in tobacco protoplasts. The transformed tobacco cells make a neomycin phosphotransferase with a transit peptide. Some of this protein is then transported into chloroplasts and processed just like the small subunit. This experiment shows that the transit peptide by itself contains sufficient information to guide a protein into plastids.

Our extensive discussions of vector DNAs should not cause us to ignore the possibility of inducing plant cells to take up and integrate foreign genes directly, that is, without a vector. There have been many attempts to demonstrate such an event. Some successes have been reported. For example, Potrykus and his colleagues at Frederich Meischer Institute have induced cells of various plant species to incorporate an *E. coli* gene directly, using antibody selection techniques to prove that the gene is present and active in the plant cells.

Potrykus used a gene that provides a highly selectable trait, that is, resistance to the drug kanamycin. Without this gene, plant cells (like bacteria) are killed by the drug. The gene was attached to the front end of the CaMV ORF VI, in the hope that this might provide an effective promoter. (This attachment, however, was not shown to be important.) This hybrid DNA was then added to wall-less protoplasts of tobacco in the presence of polyethylene glycol (PEG), which stimulates protoplasts to take up exogenous materials. The protoplasts were then cultured under conditions that allowed them to grow cell walls, to start dividing to form a callus, and to develop shoots and roots. At a certain point, about a week after the DNA-PEG treatment, kanamycin was added so that only cells with the kanamycin resistance trait actually completed development.

The evidence that the regenerated tobacco plants actually contained a gene for kanamycin resistance included the following: (a) hybridization, on a Southern blot, of plant cell DNA to a radioactively-labeled kanamycin-resistance gene; (b) the presence of kanamycin phosphorylating activity, which inactivates the drug; (c) resistance of leaves to a kanamycin spray, indicated by contrast to the control leaves, which turned yellow and died; and (d) Mendelian inheritance of the resistance in the progeny of 21 out of 24 self-fertilized regenerated plants. The allele for resistance was dominant.

Potrykus's results strongly suggest that plant cells can take up and integrate a foreign gene under the right circumstances, even in the absence of a vector. The technique works with cells of monocots as well as dicots. This is important since *Agrobacterium* will not infect monocot cells and cannot be used to insert foreign genes into these plants. However, without a vector, the efficiency of transformation is not high. The original experiments had efficiencies of $10^{-4}$ to $10^{-6}$, though with optimized conditions the efficiency can be raised to $10^{-3}$. In any case, highly stringent selective pressure is especially important in locating the transformed cells. Thus, it is likely that vectors will continue to be used in future experiments when possible.

### Technical insert:  Electroporation

Most of the experiments on transformation reported in Section D depended on the triparental (pMON) *Agrobacter* system for injecting new genes into cells. Sometimes this system is not satisfactory. For instance, *Agrobacter* does not infect monocots. Electroporation, a technique using electrical fields to make protoplasts temporarily permeable to DNA, offers an effective alternative.

In electroporation, plant cell protoplasts are suspended in a small chamber that has electrodes at opposite ends. The suspension medium contains the DNA or other material to be inserted into the cells. Pulses of high voltage are applied to the electrodes. The high voltage produces pores in the plasma membranes, allowing the DNA (or other material) to diffuse into the cells. After a short period, the membranes reseal. If properly treated, the cells can then regenerate cell walls, divide to form a callus, and finally regenerate new plants.

The critical part of this procedure is finding and establishing conditions that produce pores that are sufficiently large and stay open long enough to allow the DNA insertion. Yet these same conditions must make pores that are reversible and temporary. (See Figure 6–15.) Reversible pores are induced by a potential of about 1 volt across a membrane, but the exact conditions that will induce this voltage and make it work correctly depend on the size of the cell, the composition of the cell membrane, the composition of the suspending solution, and the temperature. Applying a $40$-$\mu$sec pulse of electric field at 2 kV/cm to red blood cells causes an exchange of sodium and potassium across their plasma membranes. The pores that allow this exchange reseal in seconds or fractions of a second at normal temperatures (37°C), but they stay open for several minutes at 0°C. Increasing the field strength slightly allows an exchange of larger molecules. Increasing it to 3 kV/cm irreversibly lyses the cells unless stabilizing agents (divalent anions, EDTA, inulin, or sucrose) are added to the medium.

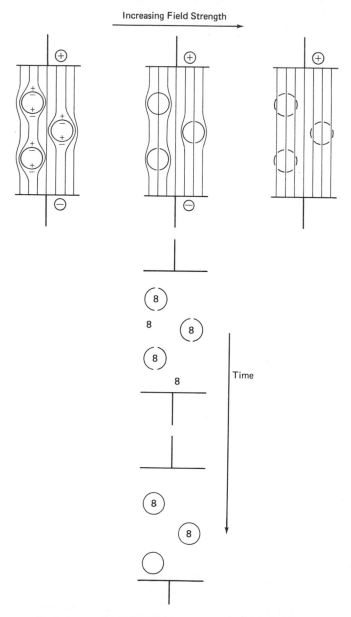

**Figure 6-15.** Electroporation of protoplasts. The top line shows the effect of increasing field strength. At low field strength, the field and current do not penetrate the cells (though the field does induce voltages across the cell membranes). At intermediate field strength, reversible pores open in the membranes, and current crosses the cells. At high field strength, very many pores are opened, and irreversible lysis occurs. The vertical sequence indicates that cells treated with the intermediate field strength are able to take up molecules of DNA in the medium before their pores reseal.

It has been possible to insert a "cat" (chloramphenicol acetyltrans-ferase) gene attached to a cauliflower mosaic virus promoter into rice, wheat, and sorghum protoplasts using electroporation. Ou-Lee and co-workers in 1986 treated approximately 400,000 protoplasts and 4 $\mu$g of plasmid DNA containing the cat gene with two 100 $\mu$sec pulses at 2.5 kV/cm. (These are representative conditions; Ou-Lee and his associates tried several variations to get optimum results.) They held the protoplasts at 0°C for ten minutes, then placed the protoplasts in culture medium. Two days later, they detected cat activity in extracts of the protoplasts.

Electroporation thus may become the method of choice for inserting DNA into plant cells. Getting this DNA to incorporate efficiently into the plant genome, and especially into specified places in the plant genome, will probably still require vectors as described earlier in this chapter.

# SEVEN

# Summary

Plant development involves a complex series of morphological, anatomical, and biochemical events. Control of these events has remained mysterious, although it is clear that the development of a given part of a plant is influenced by internal factors such as the present state of plant development, the presence of adjacent tissues, and the environmental factors of light, temperature, day length, and so forth. There are a few major growth regulators that are known to initiate and direct the developmental processes, though it is not known how these function. Nor is it known why only some steps in development are clearly affected by hormones, or why a given hormone has different effects on different tissues, or why a given developmental step is triggered by different hormones in different species of plants.

A study of plant molecular genetics may be expected to help us understand plant development in at least two ways. First, gene regulation is a part of the differentiation of many tissues. Gene regulation is a biochemical process that can eventually be unraveled at the molecular level. When we understand how genes are controlled in development, we will know in part how development is controlled. Second, certain genes control development, in that their different alleles change the timing or direction of certain developmental processes. Learning how these genes work will teach us about the internal controls to development, which now seem so obscure.

In investigating plant genes and how they act, we can form hypotheses from our extensive knowledge of bacterial and animal molecular genetics. There are several systems in bacteria by which gene expression is turned off or on, and the principal regulatory steps and factors are known. In general, transcription is the most common step to be regulated. The regulation involves repressors, as in the lactose and tryptophan operons, activators, such as the CAP protein in the lactose operon, or sigma factors.

There are also several systems in animal cells in which gene control has been extensively studied. From these studies, it is apparent that the genes and gene functions in eukaryotic nuclei may be more complex than genes and gene functions in prokaryotes. Active gene expression is associated with changes in chromatin structure, as in the puffs of insect chromosomes. To stimulate gene activation, there may be a change in the physical structure of DNA, such as demethylation or the formation of Z-DNA. In certain unusual cases, gene activation even involves a reorganization of DNA base sequences, as with immunoglobulin genes. The RNA that is transcribed from active DNA templates (hnRNA or nRNA) contains introns that must be removed; other processing steps are also required to form a competent mRNA. Translation of the mRNA forms a polypeptide chain that in many cases must also be processed before it becomes a functional enzyme. Each of these steps—activation, transcription, RNA processing, translation, polypeptide chain processing—must be performed without error for a gene to be expressed. To learn how a given gene is controlled during development, we must first learn which step or steps are not performed in cells that are not expressing that gene.

## A. HOW GENES ARE CONTROLLED IN DEVELOPMENT

In the several plant systems that we described, the induction of a protein—the result of a change in the control of expression of that protein's gene—correlated with, and probably depended upon, an increase in mRNA. This was true for PAL and other enzymes of the phenolic pathway, alcohol dehydrogenase, auxin-stimulated proteins, seed proteins, the small subunit of RuBP carboxylase, and the light-harvesting chloroplast protein. In those cases that were tested, it appeared that transcription of the appropriate genes had increased. With the seed proteins, there was evidence that changes in post-transcriptional processes, the rate of processing of the nRNA, or the rate of breakdown of nRNA or mRNA, might have a strong influence on the amount of mRNA that accumulated. In one situation, the large subunit of RuBP carboxylase in plants treated with cycloheximide, the lack of protein was ascribed to rapid degradation rather than a lack of synthesis. Thus there were several ways in which gene expression was controlled, though the emphasis was definitely on gene activation and transcription.

Note that the genes that have been studied are in a special class. They are genes that produce a large amount of protein or which produce protein very rapidly under particular conditions. These are just those characteristics that make the genes easy to study. But it is possible that these same characteristics may require a particular mechanism of regulation. If this is true, then the generalizations we draw from these studies may not apply to other classes of genes.

Whereas the observations highlight locations at which gene regulation occurs, they do not yet illuminate the mechanisms of control. We must learn how hormones, phytochrome, and nutritional supply and balance, among other factors, influence gene expression. Do these act directly, through DNA-binding or RNA-binding receptor proteins, or do they act indirectly, through modulation of cyclic AMP or calcium concentrations, for instance? If they act indirectly, what agent actually acts on the nucleic acids? What features of the genome allow specific genes to be recognized and turned on or off independently of other genes?

The study of the plant nuclear genome reveals all the complexities of animal genomes. Some of these are even exaggerated. There are often large numbers of repeated sequences interspersed widely through the coding regions. Many of these repeated sequences are transcribed but not processed or translated. What is their function? Does it include marking genes for regulatory purposes? Though this idea is attractive, there is little or no evidence in plant systems to support it, and the wide variation in repeated DNA among closely related species argues against an important regulatory role for the repeated sequences.

The other plant cell genomes—plastid and mitochondrial—are distinctly different from the nuclear genome, and their genes are probably controlled differently. The organization of the plastid genome is more bacterial-like. The plastid genome has at least one polygenic mRNA, but other features of operons (repressors and operators) have not yet been detected. The mitochondrial genome is more mysterious, and it is difficult even to speculate about the types of control mechanisms at work there.

## B. HOW GENES CONTROL DEVELOPMENT

The literature of classical plant genetics provides many examples of genes that direct the course of development. These include genes that affect the size, color, or position of organs like leaves and flowers. A few genes have been identified as structural genes for particular proteins, but the majority have not been identified specifically. Most of these latter genes must be regulatory genes that control the timing or degree of expression of structural genes.

It is curious that, despite the prevalence of regulatory genes in the catalogue of plant geneticists, we know so little about how these genes act. In the past it was difficult to study them at the molecular level because their products (RNA or protein) were not known. But now new techniques for the modification of genes allow the formation of mutant alleles with inserted sequences that can be detected in RNA transcripts. These techniques allow workers to identify the primary gene products of the mutated genes and, by inference, the products of the wild type genes. These techniques may also give clues to the function of the genes by showing exactly what changes in base sequence disturb their functions.

The tumorigenic genes of the Ti plasmid demonstrate another way in which studies of gene modification lead to an understanding of the genetic control of development. Tumorigenic genes have effects that clearly redirect development, inducing callus, or shoots, or roots in situations where these organs would not normally appear. It is a valuable step to link these genes to the production of well-known hormones (for instance, the tmr gene to isopentenyl transferase), for it confirms predictions made on the basis of indirect evidence. Such a connection disproves earlier speculations about less well defined mechanisms that would be independent of hormones. From these studies, we can expect more information about the genetic control of hormone concentrations in plant cells.

As described at the beginning of Chapter 6, we can predict that in the future regulatory genes will be studied by inserting them into new cells or into new loci of chromosomes. An approach like this can be very sophisticated, allowing an investigator to combine many single genes in order to show the interactions among them and to study the influence of other structural aspects of DNA on gene expression.

Plant development is a broad and complex field, and its problems will not all be resolved in the near future. But the techniques of molecular biology will continue to provide new insights into some of the basic questions that have puzzled developmental botanists for years.

# Bibliography

## GENERAL REFERENCES

CIFERRI, O., AND L. DURE III. *Structure and Function of Plant Genomes.* NATO Advanced Science Institutes Series, Vol. 63. Plenum, New York, 1983.

GOLDBERG, R. *Plant Molecular Biology.* UCLA Symposium on Molecular Biology, New Series, Vol. 12. Alan R. Liss, New York, 1983.

KOSUGE, T., C. MEREDITH, AND A. HOLLAENDER. *Genetic Engineering of Plants: An Agricultural Perspective.* Plenum, New York, 1983.

LEWIN, B. *GENES II.* John Wiley & Sons, New York, 1985.

PARTHIER, B., AND D. BOULTER. *Nucleic Acids and Proteins in Plants, Vol. II: Structure, Biochemistry, and Physiology of Nucleic Acids.* Encyclopedia of Plant Physiology, New Series, Vol. 14, Part B. Springer-Verlag, New York, 1982.

SMITH, H. AND D. GRIERSON. *The Molecular Biology of Plant Development.* University of California Press, Berkeley, Calif., 1982.


## REFERENCES FOR CHAPTER 1

KAMALAY, J.C., AND R.B. GOLDBERG. "Organ-specific nuclear RNAs in tobacco." *Proc. Natl. Acad. Sci.* USA 81:2801–2805, 1984.

SUSSEX, I.M. "Determination of plant organs and cells." in *Genetic Engineering of Plants: An Agricultural Perspective.* T. Kosuge, C. Meredith, A. Hollaender, eds. Plenum, New York, 1983.

TREWAVAS, A.J. "Possible control points in plant development." in *The Molecular Biology of Plant Development.* Chap. 2. Plenum, New York, 1983.

TREWAVAS, A.J. "Growth substance sensitivity: the limiting factor in plant development." *Physiologia Plantarum* 55:60–72, 1982.

WAREING, P.F. "Determination and related aspects of plant development." in *The Molecular Biology of Plant Development*. H. Smith and D. Grierson, eds. Chap. 19. University of California Press, Berkeley, Calif., 1982.


## REFERENCES FOR CHAPTER 2

BROWN, D.D. "Gene expression in eukaryotes." *Science* 211:667–674, 1981.

GALLING, G. "Use (and misuse) of inhibitors in gene expression." in *Nucleic Acids and Proteins in Plants*. Vol. II: *Structure, Biochemistry and Physiology of Nucleic Acids. Encyclopedia of Plant Physiology, New Series, Vol. 14, Part 3*. B. Parthier and D. Boulter, eds. Springer-Verlag, New York, 1982.


## REFERENCES FOR CHAPTER 3, SECTION A: HEAT SHOCK

ASHBURNER, M. "The effects of heat shock and other stress on gene activity: an introduction." in *Heat Shock: From Bacteria to Man*. M.J. Schlesinger, A. Tissieres, M. Ashburner, eds. Cold Spring Harbor Laboratory, Cold Spring Harbor, N. Y., 1982.

COOPER, P., AND T-H.D. HO. "Heat shock proteins in maize." *Plant Physiology* 71:215–222, 1983.

KEY, J.L., C.Y. LIN, AND Y.M. CHEN. "Heat shock proteins of higher plants." *Proc. Natl. Acad. Sci.* USA 78:3526–3530, 1981.

LAEMMLI, U.K. "Cleavage of structural proteins during the assembly of the head of bacteriophage T4." *Nature* 227:680–685, 1970.

MEYER, Y., AND Y. CHARTIER. "Long-lived and short-lived heat-shock proteins in tobacco mesophyll protoplasts." *Plant Physiology* 72:26–32, 1983.

O'FARRELL, P.H. "High resolution two-dimensional electrophoresis of proteins." *Jour. Biological Chemistry* 250:4007–4021, 1975.

TISSIERES, A. "Summary." in *Heat Shock: From Bacteria to Man*. M.J. Schlesinger, A. Tissieres, M. Ashburner, eds. Cold Spring Harbor Laboratory, Cold Spring Harbor, N.Y., 1982.


## REFERENCES FOR CHAPTER 3, SECTION B: PHENOLIC PATHWAY

HAHLBROCK, K. "Regulatory aspects of phenylpropanoid biosynthesis in cell cultures." in *Plant Tissue Culture and Its Bio-technological Application*. W. Barz, E. Reinhard, M.H. Zenk, eds. Springer-Verlag, New York, 1977.

HAHLBROCK, K., C.J. LAMB, C. PURWIN, J. EBEL, E. FAUTZ, AND E. SCHAFER. "Rapid response of suspension-cultured parsley cells to the elicitor from *Phytophthora megasperma* var *sojae.*" *Plant Physiology* 67:768–773, 1981.

LAWTON, M.A., R.A. DIXON, K. HAHLBROCK, AND C.J. LAMB. "Elicitor induction of mRNA activity: rapid effects of elicitor on phenylalanine ammonia-lyase and chalcone synthase mRNA activities in bean cells." *European Jour. Biochemistry* 130:131–139, 1983.

## REFERENCES FOR CHAPTER 3 SECTION C: ANAEROBIOSIS

GERLACH, W.L., A.J. PRYOR, E.S. DENNIS, R.J. FERL, M.M. SACHS, AND W.J. PEACOCK. "cDNA cloning and induction of the alcohol dehydrogenase gene (Adh-1) of maize." *Proc. Natl. Acad. Sci.* USA 79:2981–2985, 1982.

JOHNS, M.A., M. ALLEMAN, M. FREELING. "Differential regulation of the Adh-1 gene in maize: facts and theories." in *Genetic Engineering of Plants: An Agricultural Perspective.* T. Kosuge, C. Meredith, A. Hollaender, eds. Plenum, New York, 1983.

STOMMER, J.N., S. HAKE, J. BENNETZEN, W.C. TAYLOR, AND M. FREELING. "Regulatory mutants of the maize Adh-1 gene caused by DNA insertions." *Nature* 300:542–544, 1982.

## REFERENCES FOR CHAPTER 3, SECTION D: IAA-INDUCED GENE EXPRESSION

BAULCOMBE, D., J. GIORGINI, AND J.L. KEY. "The effect of auxin on the polyadenylated RNA of soybean hypocotyls." in *Genetic Engineering of Plants: An Agricultural Perspective.* T. Kosuge, C. Meredith, A. Hollaender, eds. Plenum, New York, 1983.

THEOLOGIS, A., AND P.M. RAY. "Early auxin-regulated polyadenylated mRNA sequences in pea stem tissue." *Proc. Natl. Acad. Sci.* USA 79:418–421, 1982.

THEOLOGIS, A., T.V. HUYNH, AND R.W. DAVIS. "Rapid induction of specific mRNAs by auxin in pea epicotyl tissue." *Jour. Molecular Biology* 183:53–68, 1985.

VANDERHOEF, L.N., AND R.R. DUTE. "Auxin-regulated wall loosening and sustained growth in elongation." *Plant Physiology* 67:146–149, 1981.

WALKER, J.C., AND J.L. KEY. "Isolation of cloned cDNAs to auxin-responsive poly(A)+ RNAs of elongating soybean hypocotyl." *Proc. Natl. Acad. Sci.* USA 79:7185–7189, 1982.

ZURFLUH, L.L., AND T.J. GUILFOYLE. "Auxin-induced changes in the patterns of protein synthesis in soybean hypocotyl." *Proc. Natl. Acad. Sci.* USA 77:357–361, 1980.

## REFERENCES FOR CHAPTER 3, SECTION E: SEED PROTEINS

BEACH, L.R., D. SPENCER, P.J. RANDALL, AND T.J.V. HIGGINS. "Transcriptional and post-transcriptional regulation of storage protein gene expression in sulfur-deficient pea seeds." *Nucleic Acids Res.* 13:999–1013, 1985.

CHANDLER, P.M., T.J.V. HIGGINS, P.J. RANDALL, AND D. SPENCER. "Regulation of legumin levels in developing pea seeds under conditions of sulfur deficiency." *Plant Physiology* 71:47–54, 1983.

CHAPPELL, J., AND M.J. CHRISPEELS. "Transcriptional and posttranscriptional control of phaseolin and phytohemagglutinin gene expression in developing cotyledons of *Phaseolus vulgaris*." *Plant Physiology* 81:50–54, 1986.

CHRISPEELS, M.J. "Biosynthesis, processing and transport of storage proteins and lectins in cotyledons of developing legume seeds." *Philosophical Transactions Royal Society of London* Series B 304:309–322, 1984.

GOLDBERG, R.B., G. HOSCHEK, S.H. TAM, G.S. DITTA, AND R.W. BREIDENBACH. "Abundance, diversity, and regulation of mRNA sequence sets in soybean embryogenesis." *Developmental Biology* 83:201–217, 1981.

GREENE, F.C. "Expression of storage protein genes in developing wheat (*Triticum aestivum* L.) seeds." *Plant Physiology* 71:40–46, 1983.

LARKINS, B.A. "Genetic engineering of seed storage proteins." in *Genetic Engineering of Plants: An Agricultural Perspective*. T. Kosuge, C. Meredith, A. Hollaender, eds. Plenum, New York, 1983.

WALBURG, G., AND B.A. LARKINS. "Oat seed globulins: subunit characterization and demonstration of its synthesis as a precursor." *Plant Physiology* 72:161–165, 1983.

## REFERENCES FOR CHAPTER 3, SECTION F: ROOT NODULES AND NITROGEN FIXATION

DAZZO, F.B., AND A.E. GARDIOL. "Host specificity in *Rhizobium*-legume interactions." in *Genes Involved in Microbe-Plant Interactions*. D.P.S. Verma, Th. Hohn, eds. Springer-Verlag, Wien, Austria, 1984.

FULLER, F., P.W. KUNSTER, T. NGUYEN, AND D.P.S. VERMA. "Soybean nodulin genes: analysis of cDNA clones reveals several major tissue-specific sequences in nitrogen-fixing root nodules." *Proc. Natl. Acad. Sci.* USA 80:2594–2598, 1983.

KATINAKIS, P., AND D.P.S. VERMA. "Nodulin-24 gene of soybean codes for a peptide of the peribacteroid membrane and was generated by tandem duplication of a sequence resembling an insertion element." *Proc. Natl. Acad. Sci.* USA 82:4157–4161, 1985.

MAURO, V.P., T. NGUYEN, P. KATINAKIS, AND D.P.S. VERMA. "Primary structure of the soybean nodulin-23 gene and potentially regulatory elements in the 5'- flanking regions of nodulin and leghemoglobin genes." *Nucleic Acids Res.* 13:239–249, 1985.

MIFLIN, B.J., AND J. CULLIMORE. "Nitrogen assimilation in the legume-*Rhizobium*-symbiosis: a joint endeavour." in *Genes Involved in Microbe-Plant Interactions.* D.P.S. Verma, Th. Hohn, eds. Springer-Verlag, Wien, 1984.

NGUYEN, T., M. ZELECHOWSKA, V. FOSTER, H. BERGMANN, AND D.P.S. VERMA. "Primary structure of the soybean nodulin-35 gene encoding uricase II localized in the peroxisomes of uninfected cells of nodules." *Proc. Natl. Acad. Sci.* USA 82:5040–5044, 1985.

VERMA, D.P.S., AND K. NADLER. "Legume-*Rhizobium*-symbiosis: host point of view." in *Genes Involved in Microbe-Plant Interactions.* D.P.S. Verma, Th. Hohn, eds. Springer-Verlag, Wien, Austria, 1984.

## REFERENCES FOR CHAPTER 3, SECTION G: CHLOROPLAST PROTEINS

APEL, K. "Phytochrome-induced appearance of mRNA activity for the apoprotein of the light-harvesting chlorophyll a/b protein of barley (*Hordeum vulgare*)." *European Jour. Biochemistry* 97:183–188, 1979.

BARRACLOUGH, R., AND R.J. ELLIS. "The biosynthesis of ribulose bisphosphate carboxylase: uncoupling of the synthesis of the large and small subunits in isolated soybean leaf cells." *European Jour. Biochemistry* 94:165–177, 1979.

BROGLIE, R., G. BELLEMARE, S.G. BARTLETT, N.H. CHUA, AND A.R. CASHMORE. "Cloned DNA sequences complementary to mRNAs encoding precursors to the small subunit of ribulose-1,5–bisphosphate carboxylase and a chlorophyll a/b binding polypeptide." *Proc. Natl. Acad. Sci.* USA 78:7304–7308, 1981.

CHUA, N.H., AND G.W. SCHMIDT. "Post-translational transport into intact chloroplasts of a precursor to the small subunit of ribulose-1,5–bisphosphate carboxylase." *Proc. Natl. Acad. Sci.* USA 75:6110–6114, 1978.

CHUA, N.H., AND G.W. SCHMIDT. "Transport of proteins into mitochondria and chloroplasts (Review)." *Jour. Cell Biology* 81:461–483, 1979.

HIGHFIELD, P.E., AND R.J. ELLIS. "Synthesis and transport of the small subunit of chloroplast ribulose bisphosphate carboxylase." *Nature* 271:420–424, 1978.

LINK, G. "Phytochrome control of plastid mRNA in mustard (*Sinapis alba* L.)." *Planta* 154:81–86, 1982.

MISHKIND, M.L., AND G.W. SCHMIDT. "Posttranscription regulation of ribulose-1,5–bisphosphate carboxylase small subunit accumulation in *Chlamydomonas reinhardtii*." *Plant Physiology* 72:847–854, 1983.

MOSINGER, E., A. BATSCHAUER, E. SCHAFER, AND K. APEL. "Phytochrome control of *in vitro* transcription of specific genes in isolated nuclei from barley (*Hordeum vulgare*)." *European Jour. Biochemistry* 147:137–142, 1985.

NIVISON, H.T., AND C.R. STOCKING. "Ribulose bisphosphate carboxylase synthesis in barley leaves. A developmental approach to the question of coordinated synthesis." *Plant Physiology* 73:906–911, 1983.

SILVERTHORNE, J., AND E.M. TOBIN. "Demonstration of transcriptional regulation of specific genes by phytochrome action." *Proc. Natl. Acad. Sci.* USA 81:1112–1116, 1984.

STIEKEMA, W.J., C.F. WIMPEE, J. SILVERTHORNE, AND E.M. TOBIN. "Phytochrome control of the expression of two nuclear genes encoding chloroplast proteins in *Lemna gibba* L. G-3." *Plant Physiology* 72:717–724, 1983.

THOMPSON, W.F., M. EVERETT, N.O. POLANS, AND R.A. JORGENSEN. "Phytochrome control of RNA levels in developing pea and mung-bean leaves." *Planta* 158:487–500, 1983.

TOBIN, E.M. "Light regulation of specific mRNA species in *Lemna gibba* L. G-3." *Proc Natl. Acad. Sci.* USA 75:4749–4753, 1978.

## REFERENCES FOR CHAPTER 4

BEDBROOK, J.R., M. O'DELL, AND R.B. FLAVELL. "Amplification of rearranged repeated DNA sequences in cereal plants." *Nature* 288:133–137, 1980.

BEDBROOK, J.R., J. JONES, M. O'DELL, R.D. THOMPSON, AND R.B. FLAVELL. "A molecular description of telomeric heterochromatin in *Secale* species." *Cell* 19:545–560, 1980.

BIRNSTIEL, M.L., M. BUSSLINGER, AND K. STRUB. "Transcription termination and 3' processing: the end is in site." *Cell* 41:349–359, 1985.

BROGLIE, R., G. CORUZZI, G. LAMPPA, B. KEITH, AND N-H. CHUA. "Structural analysis of nuclear genes coding for the precursor to the small subunit of wheat ribulose-1,5–bisphosphate carboxylase." *Bio/Technology* 1:55–61, 1983.

BROWN, G.G., J.S. LEE, N. BRISSON, AND D.P.S. VERMA. "The evolution of a plant globin gene family." *Jour. Molecular Evolution* 21:19–32, 1984.

CECH, T.R. "RNA splicing: three themes with variations." *Cell* 34:713–716, 1983.

DEAN, C., P. VAN DEN ELZEN, S. TAMAKI, P. DUNSMUIR, AND J. BEDBROOK. "Linkage and homology analysis divides the eight genes for the small subunit of petunia ribulose 1,5–bisphosphate carboxylase into three gene families." *Proc. Natl. Acad. Sci.* USA 82:4964–4968, 1985.

DOVER, G.A. AND R.B. FLAVELL. "Molecular coevolution: DNA divergence and the maintenance of function." *Cell* 38:622–623, 1984.

FISCHER, R.L., AND R.B. GOLDBERG. "Structure and flanking regions of soybean seed protein genes." *Cell* 29:651–660, 1982.

FLAVELL, R.B. "The molecular characterization and organization of plant chromosomal DNA sequences." *Annu. Rev. Plant Physiology* 31:569–596, 1980.

FLAVELL, R.B. "Repeated sequences and genome architecture." in *Structure and Function of Plant Genomes*. O. Ciferri, L. Dure III, eds. Plenum, New York, 1983.

FLAVELL, R.B., J. BEDBROOK, J. JONES, M. O'DELL, W.L. GERLACH, T.A. DYER, AND R.D. THOMPSON. "Molecular events in the evolution of cereal chromosomes." *Proc. Fourth John Innes Symposium,* 1979.

FLAVELL, R.B., M. O'DELL, D.B. SMITH, AND W.F. THOMPSON. "Chromosome architecture: the distribution of recombination sites, the structure of ribosomal DNA loci, and the multiplicity of sequences containing inverted repeats," in *Molecular Form and Function of the Plant Genome*. L. van Vloten-Doting, G.S.P. Groot, T.C. Hall, eds. Plenum, New York, 1985.

GEHRING, W.J. "The homeo box: a key to the understanding of development?" *Cell* 40:3–5, 1985.

KHOURY, G., AND P. GRUSS. "Enhancer elements." *Cell* 33:313–314, 1983.

LLEWELLYN, D., E. DENNIS, AND W.J. PEACOCK. "The alcohol dehydrogenase genes of maize: expression studies in tobacco cells." in *Molecular Form and Function of the Plant Genome*. L. van Vloten-Doting, G.S.P. Groot, T.C. Hall, eds. Plenum, New York, 1985.

MESSING, J., D. GERAGHTY, G. HEIDECKER, N.-T. HU, J. KRIDL, AND I. RUBENSTEIN. "Plant gene structure." in *Genetic Engineering of Plants: An Agricultural Perspective*. T. Kosuge, C. Meredith, A. Hollaender, eds. Plenum, New York, 1983.

MURRAY, M.G., D.L. PETERS, AND W.F. THOMPSON. "Ancient repeated sequences in the pea and mung bean genomes and implications for genome evolution." *Jour. Molecular Evolution* 17:31–42, 1981.

MURRAY, N.E. "Phage lambda and molecular cloning." in *Lambda II*. R.W. Hendrix, J.W. Roberts, F.W.Stahl, R.A. Weisberg, eds. Cold Spring Harbor Laboratory, Cold Spring Harbor, N.Y., 1983.

PARKER, C.S. AND J. TOPOL. "A *Drosophila* RNA polymerase II transcription factor binds to the regulatory site of an hsp 70 gene." *Cell* 37:273–283, 1984.

ROBERTS, J.M., L.B. BUCK, AND R. AXEL. "A structure for amplified DNA." *Cell* 33:53–63, 1983.

SCHIMKE, R.T. "Gene amplification in cultured animal cells." *Cell* 37:705–713, 1984.

SHEPHERD, N.S., Z. SCHWARZ-SOMMER, J.B. VEL SPALVE, M. GUPTA, U. WIENAND, AND H. SAEDLER. "Similarity of the Cin1 repetitive family of *Zea mays* to eukaryotic transposable elements." *Nature* 307:185–187, 1984.

SMITH, G.P. "Evolution of repeated DNA sequences by unequal cross-over." *Science* 191:528–535, 1976.

STARK, G.R. "Gene amplification." *Annu. Rev. Biochemistry* 53:447–491, 1984.

SUTCLIFFE, J.G., R.J. MILNER, J.M. GOTTESFELD, AND W. REYNOLDS. "Control of neuronal gene expression." *Science* 225:1308–1315, 1984.

THOMPSON, W.F. AND M.G. MURRAY. "The nuclear genome: structure and function." *The Biochemistry of Plants* 6:1–81, 1981.

THOMPSON, W.F., M.G. MURRAY, AND R.E. CUELLAR. "Contrasting patterns of DNA sequence organization in plants." in *Genome Organization and Expression in Plants*. C.J. Leaver, ed. Plenum, New York, 1980.

VAN VLOTEN-DOTING, L., G.S.P. GROOT, AND T.C. HALL, EDS. *Molecular Form and Function of the Plant Genome*. Plenum, New York, 1985.

WIMPEE, C.F., W.J. STIEKEMA, AND E.M. TOBIN. "Sequence heterogeneity in the RuBP carboxylase small subunit gene family of *Lemna gibba*." *Plant Molecular Biology* 2:391–401, 1983.

## REFERENCES FOR CHAPTER 5, SECTION A: CHLOROPLAST GENOME STRUCTURE

BOGORAD, L. "Evolution of organelles and eukaryotic genomes." *Science* 188:891–898, 1975.

BOGORAD, L. "Regulation of intracellular gene flow in the evolution of eukaryotic genomes." in *On the Origin of Chloroplasts.* J.A. Schiff, ed. Elsevier North Holland, 1982.

BOHNERT, H.J., E.J. CROUSE, AND J.M. SCHMITT. "Organization and expression of chloroplast genes." *Encyclopedia of Plant Physiology* 14B:475–530, 1982.

GRUISSEM, W., AND G. ZURAWSKI. "Identification and mutational analysis of the promoter for a spinach chloroplast transfer RNA gene." *EMBO Jour.* 4:1637–1644, 1985.

HALLICK, R.B., AND W. BOTTOMLEY. "Proposals for the naming of chloroplast genes." *Plant Molecular Biology Reporter* 1:38–43, 1982.

HERRMANN, R.G., O. WESTHOFF, J. ALT, P. WINTER, J. TITTGEN, C. BISANZ, B.B. SEARS, N. NELSON, E. HURT, G. HAUSKA, A. VIEBROCK, AND W. SEBALD. "Identification and characterization of genes for polypeptides of the thylakoid membrane." in *Structure and Function of Plant Genomes.* O. Ciferri, L. Dure III, eds. Plenum, New York, 1983.

HIRSCHBERG, J., AND L. MCINTOSH. "Molecular basis of herbicide resistance in *Amaranthus hybridus.*" *Science* 222:1346–1349, 1983.

KOLLER, B., AND H. DELIUS. "Intervening sequences in chloroplast genomes." *Cell* 36:613–622, 1984.

KREBBERS, E.T., I.M. LARRINUA, L. MCINTOSH, AND L. BOGORAD. "The maize chloroplast genes for the beta and epsilon subunits of the photosynthetic coupling factor CF1." *Nucleic Acids Research* 10:4985–5002, 1982.

OHYAMA, K., H. FUKUZAWA, T. KOCHI, H. SHIRAI, T. SANO, S. SANO, K. UMESONO, Y. SHIKI, M. TAKEUCHI, Z. CHANG, S. AOTA, H. INOKUCHI, AND H. OZEKI. "Chloroplast gene organization deduced from complete sequences of liverwort *Marchantia polymorpha* chloroplast DNA." *Nature* 322:572–574, 1986.

PALMER, J. D. "Comparative organization of chloroplast genomes." *Annu. Rev. Genetics* 19:325–354, 1985.

PALMER, J.D., AND W.F. THOMPSON. "Chloroplast DNA rearrangements are more frequent when a large inverted repeat is lost." *Cell* 29:537–550, 1982.

POLANS, N.O., N.F. WEEDEN, AND W.F. THOMPSON. "Inheritance, organization and mapping of *rbcS* and *cab* multigene families in pea." *Proc. Natl. Acad. Sci.* USA 82:5083–5087, 1985.

WHITFELD, P.R., AND W. BOTTOMLEY. "Organization and structure of chloroplast genes." *Annu. Rev. of Plant Physiology* 34:279–310, 1983.

## REFERENCES FOR CHAPTER 5, SECTION B:
## MITOCHONDRIAL GENOME STRUCTURE

BENDICH, A. "Plant mitochondrial DNA: unusual variation on a common theme." in *Genetic Flux in Plants.* B. Hohn, E. Dennis, eds., pp. 111–138, Springer-Verlag, New York, 1985.

GRAY, M.W. "Mitochondrial genome diversity and the evolution of mitochondrial DNA." *Canadian Jour. Biochemistry* 60:157–171, 1982.

LEAVER, C.J., AND M.W. GRAY. "Mitochondrial genome organization and expression in higher plants." *Annu. Rev. of Plant Physiology* 33:373–402, 1982.

LEAVER, C.J., L.K. DIXON, E. HACK, T.D. FOX, AND A.J. DAWSON. "Mitochondrial genes and their expression in higher plants." in *Structure and Function of Plant Genomes.* O. Ciferri, L. Dure III, eds. Plenum, New York, 1983.

LONSDALE, D.M. "A review of the structure and organization of the mitochondrial genome of higher plants." *Plant Molecular Biology* 3:201–206, 1984.

LONSDALE, D.M., T.P. HODGE, AND C.M.-R. FAURON. "The physical map and organisation of the mitochondrial genome from the fertile cytoplasm of maize." *Nucleic Acids Research* 12:9249–9261, 1984.

LEVINGS, C.S. "The plant mitochondrial genome and its mutants." *Cell* 32:659–661, 1983.

PALMER, J.D., AND C.R. SHIELDS. "Tripartite structure of the *Brassica campestris* mitochondrial genome." *Nature* 307:437–440, 1984.

STERN, D.B., AND J.D. PALMER. "Extensive and widespread homologies between mitochondrial DNA and chloroplast DNA in plants." *Proc. Natl. Acad. Sci.* USA 81:1946–1950, 1984.

WALBOT, V., K.J. NEWTON, A. MALONEY, T. SANDIE, B.S. MASTERS, D. MCCARTY, E. FEJES, AND W.W. HAUSWIRTH. "Mapping genes of the maize mitochondrial genome." *International Chemical and Nuclear-University of California at Los Angeles Symposia,* 1983.

WARD, B.L., R.S. ANDERSON, AND A.J. BENDICH. "The size of the mitochondrial genome is large and variable in a family of plants (Cucurbitaceae)." *Cell* 25:793–803, 1981.

## REFERENCES FOR CHAPTER 6

BRISSON, N., J. PASZKOWSKKI, J.R. PENSWICK, B. GRONENBORN, I. POTRYKUS, AND T. HOHN. "Expression of a bacterial gene in plant by using a viral vector." *Nature* 310:511–514, 1984.

BROGLIE, R., G. CORUZZI, R.T. FRALEY, S.G. ROGERS, R.B. HORSCH, J.G. NIEDERMEYER, C.L. FINK, J.S. FLICK, AND N.-H. CHUA. "Light-regulated expression of a pea ribulose-1,5–bisphosphate carboxylase small subunit gene in transformed plant cells." *Science* 224:838–843, 1984.

CAPLAN, A., L. HERRERA-ESTRELLA, D. INZE, E. VAN HAUTE, M. VAN MONTAGU, J. SCHELL, AND P. ZAMBRYSKI. "Introduction of genetic material into plant cells." *Science* 222:815–821, 1983.

CHILTON, M.-D. "A vector for introducing new genes into plants." *Scientific American* 248(6):50–59, 1983.

DORING, H.-P., AND P. STARLINGER. "Barbara McClintock's controlling elements: now at the DNA level." *Cell* 39:253–259, 1984.

FEDOROFF, N.V. "Transposable genetic elements in maize." *Scientific American* 250(6):84–98, 1984.

FREELING, M.J. "Plant transposable elements and insertion sequences." *Annu. Rev. of Plant Physiology* 35:277–298, 1984.

HERRARA-ESTRELLA, L., G. VAN DEN BROECK, R. MAENHAUT, M. VAN MONTAGU, J. SCHELL, M. TIMKO, AND A. CASHMORE. "Light-inducible and chloroplast-associated expression of a chimaeric gene introduced into *Nicotiana tabacum* using a Ti plasmid vector." *Nature* 310:115–120, 1985.

HORSCH, R.B., R.T. FRALEY, S.G. ROGERS, P.R. SANDERS, A. LLOYD, AND N. HOFFMAN. "Inheritance of functional foreign genes in plants." *Science* 223:496–498, 1984.

HOWELL, S.H. "Plant molecular vehicles: potential vectors for introducing foreign DNA into plants." *Annu. Rev. of Plant Physiology* 33:609–650, 1982.

MARX, J.L. "Ti plasmids as gene carriers." *Science* 216:1305–1307, 1982.

MORELLI, G., F. NAGY, R.T. FRALEY, S.G. ROGERS, AND N.-H. CHUA. "A short conserved sequence is involved in the light-inducibility of a gene encoding ribulose-1,5–bisphosphate carboxylase small subunit of pea." *Nature* 315:200–204, 1985.

MULLER-NEUMANN, M., J.I. YODER, AND P. STARLINGER. "The DNA sequence of the transposable element *Ac* of *Zea mays* L." *Molecular and General Genetics* 198:19–24, 1984.

MURAI, N., D.W. SUTTON, M.G. MURRAY, J.L. SLIGHTOM, D.J. MERLO, N.A. REICHERT, C. SENGUPTA-GOPALAN, C.A. STOCK, R.F. BARKER, J.D. KEMP, AND T.C. HALL. "Phaseolin gene from bean is expressed after transfer to sunflower via tumor-inducing plasmid vectors." *Science* 222:474–482, 1983.

OU-LEE, T.-M., R. TURGEON, AND R. WU. "Uptake and expression of a foreign gene linked to either a plant virus or *Drosophila* promoter, after electroporation of protoplasts of rice, wheat and sorghum." *Proc. Natl. Acad. Sci.* USA 83:6815–6819.

POTRYKUS, I., J. PASZKOWSKI, M.W. SAUL, J. PETRUSKA, AND R.D. SHILLITO. "Molecular and general genetics of a hybrid foreign gene introduced into tobacco by direct gene transfer." *Molecular General Genetics* 199: 169–177, 1985.

REAM, L.W., AND M.P. GORDON. "Crown gall disease and prospects of genetic manipulation of plants." *Science* 218:854–859, 1982.

SCHREIER, P.H., E.A. SEFTER, J. SCHELL, AND H.J. BOHNERT. "The use of nuclear-encoded sequences to direct the light-regulated synthesis and transport of a foreign protein into plant chloroplasts." *EMBO Jour.* 4:25–32, 1985.

SHEPHERD, R.J. "DNA plant viruses." *Annu. Rev. of Plant Physiology* 30:405–423, 1979.

ZIMMERMANN, U. "Electric field-mediated fusion and related electrical phenomena." *Biochimica Biophysica Acta* 694:227–277, 1982.

# Index

## A

Abscisic acid, *17*, 20, *20*
Abscission, 7, 85
*Acetabularia*, 146
Acid-growth hypothesis, 77
Acrylamide, 58
Actinomycin D, 52, 77
Activation, gene, 41, *42*
Addition, poly(A), 130
Aerenchyma, 93
*Agrobacterium:*
  *A. rhizogenes*, 184
  *A. tumefaciens*, 184
Alcohol dehydrogenase, 65
  electrophoretic mobility of, *67*
  introns in genes of, 128
  isozymes of, *67*
Aleurone, 19
Alfalfa, 148
Allantoic acid, 93
*Alnus*, 91
Alpha-amanitin, 52
Alpha-amylase, 19
Amido schwartz, as PAGE stain, 58
Ampholytes, 59
Amplification, DNA, 45, 110, 116, *118*

*Anabena*, 91
Anaerobiosis, 65
Anderson, R., 164
Androecium, 7
Annealing, 120
Anoxia, 55
Anthocyanins, *26*, 60, 62
Apocytochrome *b*, 172
*Arabidopsis thaliana*, 119
Asparagine, 93
*Aspergillus*, 106
  *A. oryzae*, 122
ATP synthetase, 98, 152
Attenuation, 39
Autoradiography, 58, 71, 79
Auxin, 11, 15–16, *16*, *17*, 77, *78*, *80*, 185
Axel, R., 110
*Azolla*, 91

## B

*Bacillus subtilis*, 38, 153
Bacteroid, *92*, *96*
Barr Body, 40, 44
Beach, L.R., 88
Bendich, A., 164

Blot:
  Northern, 71, 75, 79, *80,* 156
  Southern, 75, *76,* 150, *161,* 162
  Western, 75, 189
Blot hybridization, 75
Bogorad, L., 152, 158
*Brassica,* 164, 165, 168
  *B. napus, 13*
Braun, A., 185
Brisson, N., 189
Britten, R., 114
Broad bean, 148, 153
Broglie, R., 194
Bundles, vascular, 4


*C*


C-value paradox, 105
Callus, 11, *13, 185*
Calyx, 7
Capping, RNA, 47
CAP protein, 36
Carpel, 7, *8*
Cauliflower, 181
Cauliflower mosaic virus, 181, *182,*
  189
  DNA of, *183*
CF0, 98
CF1, 99
*Chalcone synthetase,* 60, 64
Chappell, J., 88
Chilton, M.-D., 186
*Chlamydomonas reinhardtii,* 102, 126
Chloramphenicol, 52, 102
Chlorophyll *a/b* binding protein (*see*
  Light-harvesting chloroplast
  protein)
Chloroplast:
  DNA of, 146, 147
  genes of, 148, 151
  genes of, in mitochondria, 169
  rRNA genes of, 153
  transcript patterns in, *157*
  tRNA genes of, 156
Chrispeels, M.J., 88
Chromatin, 40, *41*
  heterochromatic, 40, 105
Chromosome, 39
Chromosome puff, 41, 55
Chromosome walking, 162, 165
Chua, N.-H., 100, 145

Cloning:
  of bacteria, 69
  of cDNA, 68, 73
  genomic, 123
Clover, 153
Coleoptile, 4
*Colletotrichum lindemunthianum,* 64
Conglycinin, 84, 87, *128*
Coomassie Brilliant Blue, as PAGE
  stain, 58, *59*
Copy number, 106
Cordycepin, 52
Corepressor, 36
Corolla, 10
Coruzzi, G., 145
Cos sites, lambda bacteriophage, 124
$C_ot$ *121,* 123
$C_ot$ curve, *121*
Cotyledon, 2, 67, 81, 86
Coumarate CoA ligase, 60
Coupling factor (*see* ATP synthetase)
Crown, 184
Crown gall disease, 184, *185*
Cryptochrome, 23
Culture, plant tissue and cell, 28
cDNA (complementary DNA, copy
  DNA), 68
  cloning of, *74*
  preparation of, *69,* 72, *73*
*Cyanophora,* 152
Cycloheximide, 52, 77, 102
Cytochrome *b559,* 98
Cytochrome *b6/f,* 98, 152
Cytokinin, 11, *17,* 19, 185
cytoplasmic male sterility, 168


*D*


*Daucus carota, 18*
Davidson, N., 114
Dean, C., 144
De-etiolation, 101
Delius, H., 158
Denaturation, DNA, 120
Determination, in development, 11,
  12
Dichlorophenoxyacetic acid (2,4-D),
  as auxin, 79
Dicot, 2
*Dictyostelium,* response to heat shock,
  54, 57

Dihydrofolic reductase, 189, *190*
Dimethylsulfate (DMS), 89
Direct Repeat, 180, *180*
DNA:
  repetitive, 106, 115
  sequencing of, 88, *90*
DNAase I, 41
DNA polymerase, Klenow fragment
  of, 89
DNA synthesis, unscheduled, *111*,
  112
Dormancy, 6, 10, 20, 25
Dormant bud, *3*
Duckweed, 101, 142
Dute, R.R., 78

Ferridoxin-NADP reductase, 98
Ferridoxin reductase, 98
Fertilization, 8
Fig, *14*
Flavell, R.B., 112, 141
Flavonoid, 25
  biochemical pathway, 60, 61
  induction of enzymes for, 62–64
Flower, *3, 8*
*Flowering,* 25
*Fluorography,* 58
*Frankia,* 91
Freeling, M., 68, 176, 177
Fruit, 9
Fusicoccin, 80


*E*

Electrophoresis:
  agarose, 71, 75
  (*see also* Polyacrylamide gel elec-
    trophoresis)
Electroporation, 175, 197, *198*
Ellis, R.J., 100
Embryo, 10
Emerson, R., 176
Endoplasmic reticulum, *49*, 81
Endosperm, 2, 9, 10, 81
Enhancer, 42, 135
Enlargement (of cells):
  isodiametric, 16, 21
  polarized, 15
Epidermis, 4
Epinasty, 15
*Escherichia coli,* 31, 34, 37, 52, 64, 69,
  75, 153, 175
  reassociation of DNA of, *117, 121,*
    123
  response to heat shock, 54
Ethidium bromide, 75, *76*
Ethylene, *14,* 15, *17,* 21
Etioplast, 4
*Euglena,* 148, 156, 158
Evolution, concerted, 140
Exon, 127
Explant, 29


*F*

Ferridoxin, 98

*G*

Gall, crown, 184
Gene:
  rearrangement of, 45
  regulatory, 26
  structural, 24
  structure of, 126, *127*
  structure of, in plants, 126, *127,*
    *131*
Gene families, for proteins, 39, 142
Gene family, for ribosomal RNA, 139
Genetic information:
  expression in eukaryotes, 39–52,
    *42*
  expression in prokaryotes, *31–39,*
    *31*
Genome:
  chloroplast, 146, *147, 154*
  chloroplast, isomerization of, *149*
  eukaryote, 39
  eukaryotic nuclear, 105
  mitochondrion, 163–64, *166, 167,*
    *172*
  prokaryote, 31
  turnover of, 116, *118*
Germination, seed, 10, 25
Gibberellic acid, *17*
Gibberellin, 18
Gilbert, W., 89, 130
*Ginkgo biloba,* 153
Glutamine synthase, 93
Glycinin, 84, 87
Glycolysis, 65
Goldberg, R.B., 86, 87

Golgi complex, 84
Gordon, M.P., 186
Ground tissue, 4
Growth:
    determinate, 7
    gravitropic, 16
    indeterminate, 4
    phototropic, 16
    primary, 5, 10, 25
    secondary, 10
Gruissem, W., 159
Guanylyl transferase, 47
Guilfoyle, T.J., 80
Gynoecium, 7

**H**

Hahlbrock, K., 62
Hair cell, root, 93
Hairpin, in RNA, *33*, 107
Harada, J., 87
Heat shock, 54, 135
Heat shock proteins, 54, *57*
Herrara-Estrella, L., 194
Heterogeneous nuclear RNA
    (hnRNA), *42*, 46, *47*, 127, 136
    processing of, 46
Histone, 40, *41*
Homeo box, 137
Hormone, 15, *17*
Horsch, R.B., 190
Hybridization *(see also* Reassociation,
    DNA):
    blot, 75
    solution, 120
Hydrazine, 91
Hydroxyapatite, 122
Hyperchromic shift, 120
Hypochromic shift, 122

**I**

Indole acetic acid, 16, *17*, 78
Inducer, 36
Induction, 14
Inositol, 11
Intron, 127, *129*, 158
Inverted repeat, 106
    chloroplast DNA, *147*, 148
    transposon, 176, 178, *179*
Isoelectric focusing, 60

Isozymes, 66, *67*

**K**

Kanamycin, 191, *192*
Katinakis, P., 95
Key, J.L., 77, 79
*Klebsiella,* 94
Koller, B., 158
Kunitz trypsin inhibitor, 87

**L**

Labeling, in vivo, 56
Lambda bacteriophage, 123, *125*
Larkins, B.A., 81, 144
Leaver, C.J., 163
Lectin, 84, 93
Leghemoglobin, 93, *96, 145*
    genes, introns in, *129*, 144, *145*
Legumin, 88
*Lemna,* 101, 142
Levings, C.S., 166
Life cycle, of flowering plants, 2, *3*
Light-harvesting chloroplast protein
    (LHCP), 98, 102
Lignin, 60
Lily, 106
Link, G., 102
Liverwort, 148
Lonsdale, D.M., 165, 168–69
Lysogen, 124

**M**

McClintock, B., 176
Macrosporophyte, *8*
Maintenance methylase, 44
Maize, 66, 81, 164, 171, 172, *172*,
    175
    endosperm cell, *82*
    heat shock in, *57*
Mapping:
    restriction, 159
    of restriction fragment length poly-
        morphisms, 143
Master circle, of mitochondrial
    genome, 166, *167*, 171
Maturation, developmental process
    before flowering, *3*, 7

Maxam (and Gilbert sequencing technique), 89
Megaspores, 7
Melting, DNA, 120
Meristem, 10, 11
  apical, 4, 10
  axillary, 6
  cork cambium, 6, 10
  definition, 4
  pericycle, 6
  root apex, 4, *5*
  shoot apex, 4, *5*
  vascular cambium, 6, 10
Messing, J., 130
Methylation, of cytosine, 43, *44*
Microspores, 7
Microsporophyte, *8*
Mitochondrion, genes of, 171
Mobilization, of storage compounds, 2, 25
M13, 89
Monocot, 2
mRNA, *42*
  mapping of the 5′ end of, *133*
Mung bean, 114, 116, *154*, 160
Murai, N., 191
Mustard, 102

## N

Neomycin phosphotransferase, 106, 190
Nester, E., 186
Nick translation, 71
*Nicotiana (see also* Tobacco):
  *N. plumbaginifolia,* 191, *192*
Nitrocellulose, 76
Nitrogen, fixation of, 91
Nitrogenase, 93
Nodule, root, 91, *92, 96*
Nodulin, 94, *96*
Nuclear RNA (nRNA), 46
Nuclease, S1, 122
Nucleoli:
  dominance hierarchy of, 142
  nucleolar organizer for, 141
Nucleosome, 40, *41,* 105

## O

Oat, 77
*Oenothera,* 172

Oligo dT, 68, 72
*Olisthodiscus,* 152
Open reading frame, 178, 183
Operator, 36
Operon, 31
  cAMP control of, *37*
  lactose (lac), 36, *37*
  polygenic, 34
  repressor control of, *37*
  tryptophan (trp), 36
Opine, 184
*Osmunda cinnamomea,* 153

## P

Palindrome, *70*
Palmer, J.D., 165, 169
*Papaver somniferum, 29*
pBR322, 73
Pea, 77, 114, 116, 142, *143,* 145, 148, 153, 154, 170
  reassociation of DNA of, *115, 117*
Peacock, W.J., 68
Peribacteroid membrane, 93, 95
Peribacteroid space, 96
Petal, 7
Petunia, 12, 142, 144, 194, *195*
Phaseolin, 191, *193*
*Phaseolus vulgaris,* 88, *92*
Phenolic pathway, 60
Phenylalanine ammonia lyase (PAL), 60
Phillips, R., 141
Phloem, 12
Photosynthesis, 97
  enzymes of, *98*
Phytoalexins, 60, 64
Phytochrome, 22, 97, 101
Phytohemagglutinin, 88
*Phytophthora megasperma,* 64
Plaques, purification of, 126
Plasmid, 69, 73
  mitochondrial, 168
  Sym, 94
  Ti, 186, *187, 195*
  Ti, as vector, 189
Pollen, 7
  of maize, *178*
Poly(A)+ RNA, 68
Polyacrylamide gel electrophoresis (PAGE), 56, 58, *59,* 81
  of maize heat shock proteins, *57*

Polyacrylamide gel electrophoresis (*cont.*)
  sodium dodecylsulfate (SDS)-, 58, 79
Polyadenylation, 47, *131, 173*
Polygenic mRNA, of CaMV, 183
Polymorphism, restriction fragment, of DNA, 143
Potrykus, I., 196
Primary growth, 4–6
Primer, DNA, 89
Pring, D., 168
Probe, 71, 76
Processing:
  phosphorylation in protein, 51
  polypeptide, 34
  protease action in protein, 34, 50–51
  protein, 49, 50, 85
  RNA, 46
Promiscuous DNA, 170
Promoter, 31, *32*
  CAAT box, 46, 135
  chloroplast, 158, *160*
  Goldberg-Hogness box, 46
  Pribnow box, 81
  TATA box, 132
Protein body, in seeds, *82, 84*
Proteins:
  antenna, 97
  chloroplast, 97
Protein synthesis, in vitro, 64
Protoplasts, *13*

**R**

Ray, P.M., 78
Reassociation, DNA (*see also* Hybridization):
  in solution, *115, 117,* 120, 121, *121*
Regulatory gene, trans-acting, 138
Replica, of viral plaques, 126
Replication, DNA, unscheduled, *111,* 112
Repression, 36, *37*
Repressor, 36, *37*
Restriction enzyme, 69, *70*
Restriction maps, 147, 159, *161*
*Rhizobium,* 91, 94
  *R. japonicum,* 93
  *R. phaseoli, 92*
  *R. trifoli,* 93

Ribonucleic acid (RNA):
  5S, 135
  ribosomal (rRNA), *140, 155*
Ribonucleic acid (RNA) polymerase:
  I, 46, 139
  II, 46
  III, 46, 135, 138, 139
  eukaryote, 46
  prokaryote, 31, *32*
  rho factor, 32
  sigma factor, 37, *38*
Ribonucleoprotein, 48
Ribulosebisphosphate carboxylase oxygenase (RuBP carboxylase, rubisco),
  large subunit, 102
  97, 99, *100,* 152
  small subunit, 99, 142, 194, *195*
Rieske protein, 99
Ritossa, F., 55
Rolling circle, model for DNA amplification, 110
Roots, adventitious, 6, *13*
Rose, K., 168
Rosette, 18
Rye, 112
  repeat sequences in DNA of, *109, 113*

**S**

Sanger, F., 88
Satellite DNA, 44, 107
Schardl, C., 168
Schell, J., 185
Schilperoort, R., 186
Sclerenchyma, 93
Scutellum, 67
Secondary growth, *3*
Seed, 2, *3,* 28
  legume, 84
  maize, 176, *177*
  proteins of, 81
Senescence, 15, 19, 22, 25
Sepal, 7
Sequence, DNA:
  identifier, 138
  single-copy, *108,* 109, 116, *117*
Sequences:
  control, 130

fossil repeat, *118*
inverted repeat, 106, *147, 154, 179*
repeat, onion skin model, *111*
repeated, 106, *107, 108,* 136, *140,*
    *166*
Sequencing, DNA, 88, *90*
Shannon, J., 77
Shearing, of DNA, 120
Shields, C., 165
Shine-Dalgarno sequence, 34
Shoots, adventitious, *13*
Signal recognition particle, 49
Signal sequence, 35, 49, 83, 95
Silver, as PAGE stain, 58, *59*
Skoog, F., 77
Smith, G.P., 110
Snapdragon, 175
Sodium dodecylsulfate (SDS), 58
Solanaceae, 12
S1 nuclease, 73, *133*
Sorghum, 168
Southern, E.M., 75
Soybean, *20,* 77, *96, 128, 145*
Spinach, *147,* 153, 170
Spliceosome, 47
Stacking gel, 58
Stamen, 7
Stern, D.B., 169, 170
Stoma, 21
Stringency, 121
Stuffer fragment, in lambda bacterio-
    phage, 124
Suberin, 6
Sugar beet, 168
Sutcliffe, J.G., 138
SV40, 42

**T**

Tandem array, of repeated se-
    quences, 107, 110
Taylor, W.C., 68, 177
T-DNA, 186, *188, 193*
*Tetrahymena,* 45, 47
    response to heat shock, 54
Theologis, A., 78, 80
Thiamine, 11
Thymidine kinase, 110
$T_m$, 121
Tobacco, 12, 48, 56, 165, 194
Tobin, E.M., 101

Tomato, *14, 26*
Totipotency, 11, 12
Transcript, polycistronic, in chloro-
    plast, 156
Transcription, 31
Transfer cells, 93
Transformation, 174, 189
    by *A. tumefaciens,* 184, 185
Transit sequence, 100, 196
Translation:
    eukaryote, 48
    prokaryote, 32, *35*
    termination, 32, *33*
Transposition, in maize, 176, *177*
Transposon, 113, 175, *177, 180*
    *Ac/Ds, 179*
    Robertson's mutator, 176

**U**

UDP apiose synthase, 64
Unequal crossover, 110

**V**

Vacuole, 84
Vanadate, 80
Vanderhoef, L.N., 77, *78*
Vector, 69, 73, 175
    expression, 126
Verma, D.P.S., 94, 95, 144
*Vicia:*
    *V. faba,* 106, 154
    *V. sativa,* 106
*Vigna radiata,* 114

**W**

Walker, J.C., 79
Walling, L., 87
Ward, B.L., 164
Wheat, 112, *140,* 164
    rye repeat sequences in, 113

**X**

X chromosome, 40
Xylem, 12

*Y*

Yeast, 54, 106, 126

*Z*

Z-DNA, 43, 136
Zeatin, *17*

Zein, 81, 130, 144
    cDNA of, *84*
    model of, *85*
    PAGE of, *83*
Zurfluh, L.L., 80
Zygote, 9, 11